The Human Farm

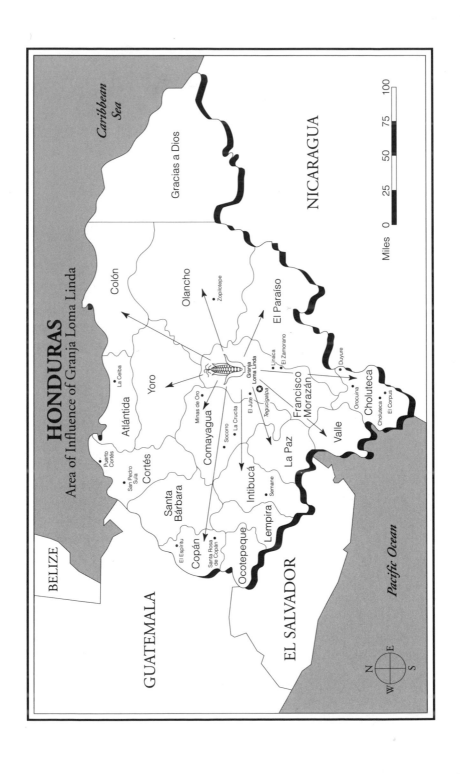

HONDURAS
Area of Influence of Granja Loma Linda

BELIZE

GUATEMALA

EL SALVADOR

Caribbean Sea

Gracias a Dios

NICARAGUA

Colón

Olancho

Zopilotepe

El Paraíso

Atlántida

La Ceiba

Yoro

Liñaca

El Zamorano

Duyure

Cortés

Puerto Cortés

San Pedro Sula

Comayagua

Minas de Oro

Socorro

La Crucita

El Jute

Tegucigalpa

Granja Loma Linda

Francisco Morazán

Orocuina

Choluteca

Choluteca

El Corpus

Santa Bárbara

El Espíritu

Copán

Santa Rosa de Copán

Intibucá

Semane

La Paz

Valle

Lempira

Ocotepeque

Pacific Ocean

Miles 0   25   50   75   100

N
W   E
S

The Human Farm

A Tale of Changing Lives and Changing Lands

Katie Smith

KUMARIAN PRESS

Kumarian Press Books for a World that Works

THE HUMAN FARM: *A Tale of Changing Lives and Changing Lands.*
Published 1994 in the United States of America by Kumarian Press, Inc.,
630 Oakwood Avenue, Suite 119, West Hartford, Connecticut 06110.

Copyright © 1994 Kumarian Press, Inc. All rights reserved.

Cover design by Laura Augustine

Production supervised by Jenna Dixon
Copyedited by Linda Lotz   *Typeset by Sarah Albert*
Proofread by Mary Frederickson   *Text design by Jenna Dixon*
Photographs by Katie Smith   *Map by Lisa Leff*
Photo layout by Kristen vonHentschel

Printed in the United States of America on recycled acid-free paper by
Thomson-Shore. Text printed with soy-based ink.

Library of Congress Cataloging-in-Publication Data

Smith, Katie, 1960–
    The human farm : a tale of changing lives and changing lands /
by Katie Smith.
       p.   cm. — (Kumarian Press books for a world that works)
    Includes bibliographical references and index.
    ISBN 1-56549-040-1 (cloth : alk. paper) — ISBN 1-56549-039-8
(paper : alk. paper)
       1. Granja Loma Linda (Honduras). 2. Farmers—Training of—
Honduras. 3. Hillside farming—Honduras. 4. Soil conservation—
Honduras. 5. Elías Sánchez, José, 1927– . 6. Flores, Milton.
7. Andrade, Fernando. I. Title. II. Series.
S539.H6G738   1994
338.1'6'097283—dc20   94-16632

98 97 96 95 94   5 4 3 2 1   1st Printing 1994

In memory of
Andrea Alejandra Flores Sandoval

He who binds to himself a joy
Doth the winged life destroy . . .

. . . But he who kisses Joy as it flies
Lives in Eternity's sunrise

—WILLIAM BLAKE

An agriculture that is whole nourishes the whole person, body and soul.

—WENDELL BERRY

Contents

Foreword

BY ROLAND BUNCH
VALLE DE ANGELES, HONDURAS

I RECENTLY READ a book called *Economic Development, The History of an Idea,* by H. W. Arndt. The book's first chapter is entitled "The Prehistory (to 1945)." Arndt's name for this chapter contains an interesting and important truth: The systematic effort to raise villager-farmers' yields and thereby contribute to their over-all well-being is a very recent human endeavor. There were, of course, isolated cases of fairly successful agricultural development work carried out before 1945. The Incas' spread of terraces and irrigation throughout much of the Andes 800 to 1,000 years ago would probably count as one of the more interesting.

Nevertheless, Arndt's point remains. Whereas the study of

ROLAND BUNCH *is founder and director of COSECHA, a Honduran-based consulting firm that has trained organizations in Africa, Asia, and Latin America in the practice of people-centered agricultural development. The methodology articulated in Mr. Bunch's seminal work* Two Ears of Corn: A Guide to People-Centered Agricultural Development *(1982) grew out of the experiences of farmers and World Neighbors trainers around the globe and Mr. Bunch's own work for World Neighbors in Central America. In Honduras, he collaborated closely with the protagonists of this book, Elías Sánchez and Milton Flores, whom he calls the pioneers of a movement that has swept Central America and parts beyond.*

mathematics began some 3,000 to 4,000 years ago, and the history of most of the subjects we study in our universities can be measured in centuries if not millennia, agricultural development as a continuous human effort basically dates back to the end of World War II. What this lack of history means, in a nutshell, is that we are still trying to discover the most basic tenets of the field. All sorts of books and articles pull development workers in contradictory directions. All too often, a promising new idea comes along and dozens of institutions jump on the bandwagon. Three or four years later, the evaluators do a few studies and find out that the new idea is not working as well as was hoped; then everyone begins looking for another new idea. On the technology side, people hailed mushrooms, fish farming, improved grains such as *triticale,* winged beans, nitrogen-fixing plants such as *leucaena*, high-lysine corn, and compost heaps as the potential saviors of vast numbers of villagers, only to learn later of their shortcomings. On the methodology side, tractorization, animal traction, cooperatives, and the transfer of technology all became tenets of programs that spread around the planet before their many shortcomings became obvious. Already, the recent enthusiasm surrounding agroforestry, integrated farms, farm-systems research, and even integrated development programs is beginning to dissipate.

Of course, some ideas that gained quick popularity have also succeeded in proving their worth in the long run; not all the roads have turned out to be dead ends. Among the ideas that have proven staying power are those that emphasize the importance of villagers using simple technologies adapted to local needs. Nevertheless, we are still searching, still trying to find our way through the vast jungle of conflicting advice, still looking for reliable signposts that will lead us to more efficient and more effective ways of helping villagers solve their many problems.

Among the worldwide pioneers in this trail-breaking effort are the two main subjects of this book: Elías Sánchez and Milton Flores. On the surface, Milton is working only with the technology of cover crops, while Elías's farm, Loma Linda, exhibits a whole wonderland of cutting-edge agricultural technologies. But as this book amply points out, both leaders are imbued with, and have been instrumental in spreading, a philosophy of how to work with village people. This philosophy, which puts transforming people at the heart of transforming rural communities and is described in this book as the "human farm," has also been incorporated into a philosophy of development increasingly known as "people-centered development." These ideas appear to be among those major ideas in agricultural development that are here to stay. Years hence, they may well form one of the bases of the general study of rural development. The ideas behind this philosophy have gradually come together over a twenty-five-year period, originating in Elías's creative mind and with the work of World Neighbors and OXFAM/ UK in Central America. That evolution is described in this book.

These ideas have spread far beyond the confines of Honduras or even Central America. Observing their surprising effectiveness, World Neighbors applied these ideas to its own programs in Mexico, Haiti, Peru, Bolivia, Togo, Burkina Faso, Indonesia, and the Philippines. With only one exception, these programs came to be seen as model programs by dozens of other institutions in their respective countries. Eventually, other development and environmental institutions, as varied as the Mennonite Central Committee, the Christian Reformed World Relief Committee, U.S. Peace Corps, CARE, Campesino A Campesino/Nicaragua, GTZ/Germany, and the World Wildlife Fund, adopted this approach for their work in one or more nations around the world.

The demand for learning about this philosophy and how to apply it to the myriad situations found in developing countries has become so great that in early 1992, a group of former World Neighbors personnel founded COSECHA, a Honduran organization dedicated largely to the dissemination of these ideas throughout Latin America and beyond. In two years, COSECHA has led workshops in twelve Latin American and two Asian nations. Successful programs using the human farm or people-centered development approach now exist in Mexico, Guatemala, Honduras, Haiti, Venezuela, Ecuador, Peru, Bolivia, Paraguay, Mali, Burkina Faso, Ghana, Togo, Kenya, India, Cambodia, Indonesia, and the Philippines.

COSECHA works with three additional institutions that have begun spreading this philosophy to other points of the globe: Trees for People/Germany, Estrategias Campesinas/Latin America, and ASPTA/Brazil.

Elías and Milton, a couple of Honduran villagers turned development workers, have shared their ideas and mixed and cross-fertilized them with those of a fair number of other villagers and development workers. These ideas have come a long way.

Preface

THE HUMAN FARM took many moons to harvest: What I first perceived as a half-year project has come to fruition two years after I set foot in Honduras in September 1992 and set eyes on the protagonists of this book. (This, you might surmise, is typical of development projects!) However, the protraction led to some pleasant personal growth. The delay required me to transport my folders, research documents, notebooks, tapes, transcriptions, computer, and ideas to four different countries on three different continents (this was not the pleasant part). As my husband and I moved from our grass-roots consulting and academic life in Arizona through a year of management studies in France, a summer of rural development work in Senegal, and family time in Canada, these various settings provided unique prisms through which to view the concepts of development espoused by the story's characters, the better and more broadly to apply them.

Along the way, I saw clearly that the principles of people-centered transformation defined in this story apply equally well on the factory floor and the family farm, in Senegal's arid Sahel and on a Honduran hillside. Indeed, the philosophy of *The Human Farm* contains general lessons in motivating healthy organizational behavior and development among communities of workers, neighbors, and kin.

One example comes readily to mind. In 1993, while working on the book, I joined a group of fellow students and faculty at the European Institute of Business Administration (INSEAD) in Fontainebleau, France, in an effort to situate business training and skills within the context of international development. Instead of trying to create a new paradigm, we tried to create new thoughts and feelings among students and faculty about the ways and meaning of enterprise. The response led to the creation of the INSEAD International Development Organization, or INDEVOR, a French not-for-profit organization that works like a greenhouse, nurturing seeds of knowledge, interest, and understanding related to sustainable development in European business students. When transplanted to the private sector, the students, not an economic model, will change the corporate landscape.

Though far removed from the rural hillsides of Honduras, such action is indeed "human farming"—changing the management of our resources by changing our minds, hearts, and souls. It is a testimony to a movement now so widespread that few know that some of its first seeds were planted by a farmer-teacher in Honduras named Don José Elías Sánchez, founder of Granja Loma Linda—a human farm.

Acknowledgments

Many helping hands reached out on the road to *The Human Farm:* first of all, those of my editor, Trish Reynolds, and publisher, Krishna Sondhi. Trish gave me her encouragement when I felt swamped, her spurs when I felt lethargic. She deserves much credit for believing in this project and bringing it to fruition.

Next, I am grateful to many friends and former colleagues around the globe with Phoenix-based Food for the Hungry International. From them I have learned, and with them I have shaped, much of my understanding of sound community development.

My parents, Mallory and Mary Ann Smith, on Bowen Island, British Columbia, graciously allowed me to hole up in seclusion during the final weeks before my manuscript deadline in January

1994 and become a part of their inspiring community. Bowen's 2,000-plus residents know and care for one another, and many of them cared for me. I think especially of the Adamses, Oslers, Allans, Cowpers, Hoopers, Judith, and Mary.

The Human Farm would not have been possible without the cooperation of its Honduran and expatriate cast of characters, whom I mention in the Introduction. But here I would like especially to thank Elías Sánchez, Candida Osorio, and the staff of Granja Loma Linda; Milton and Miriam Flores and the staff of the International Cover Crop Clearinghouse (CIDICCO); Wilmer and Miriam Dagen and the staff of World Neighbors/Honduras; and Fernando and Vilma Andrade and Roland Bunch for their generosity and openness. Milton, who first piqued my interest in the work in Honduras and then encouraged and facilitated my trip, deserves special credit for this final product.

Finally, my appreciation goes out to my husband, Michael Milway, who maintained what calm could be mustered in a hyperactive first year of marriage. He was willing on many occasions to set down his own pen and paper to listen to my prose, always offering helpful advice and loving support.

Introduction

OUT FROM the babble of politicians and the forest of paper promises, from the buzz of eco-tourists and the throb of marchers and media, I heard a clear message at the United Nations 1992 Earth Summit in Brazil: Good development starts with you, in your family, in your home, in your community. While government officials who gathered in Rio de Janeiro debated international policy, this simple message—healthy development starts at home—was carried down the city's beaches to crowds at a parallel "summit" for citizens, the Global Forum. Technology booths, speeches, music, song, dance, and religious worship paid tribute to the need for individuals and families to implicate themselves in community development in ways physical, intellectual, and spiritual.

Our meddling intellect
Misshapes the beauteous
  form of things
We murder to dissect
        —WILLIAM
        WORDSWORTH

The problem of world hunger cannot be solved until it is dealt with by local people as a multitude of local problems of ecology, agriculture, and culture.

—WENDELL BERRY
The Gift of Good Land

Some will argue that this simple message could and should have been presented without such ruckus. But the fanfare did draw attention and enthusiastic support to the most important issue of our time and of all time: our relationship to the Creator and our proper stewardship of Creation. The next questions were: Where would all the enthusiasm engendered by this international investigation-celebration-summit lead? How could it be channeled? How could it be protected from disillusionment at a political process that waters down the medicine so more will drink from the cup? What is "good development"? And what are the steps each of us could take to further it? It occurred to me that although a definition of healthy development is "both a goal and a process that seeks to achieve the broad objectives of economic equity, social justice, cultural integrity, and ecological sustainability" (see p. 125), my real bead on "good development" was much like the bead of one U.S. Supreme Court justice on "obscenity": "I know it when I see it."

The process of sustainable community development cannot be held prisoner to a phrase. We can describe steps taken to improve the use and care of resources, but such steps are never static. They are dynamic, organic, alive, not to be replicated wholly but adapted intelligently. We watch, experiment, learn, and grow. Some, like me, begin their journey of understanding with their eyes. Some, like me, learn from a good story. When I stumbled onto Granja Loma Linda in Honduras, I knew I'd found one.

The Human Farm is the story of Loma Linda—a training farm for peasant agriculturists—and its influence on a region's hillside farming families and their land. It is the story of Loma Linda's founder, Elías Sánchez, a teacher who loved his country and his kinsfolk enough to stick with them to improve their lot, even when more glamorous callings beckoned. It is the story of a peasant couple, Fernando and Vilma Andrade, bound by machismo and tradition to a way of farming that would rob their family of its meager sustenance; and of a well-educated young agronomist, Milton Flores, who turned his savvy toward environmental conservation and his method toward men's and women's hearts.

This is a true story about four people whose lives came together in pain and fear, joy and abundance in peasant villages on the hillsides of Honduras. And it is about the many others whom their lives have touched, emancipated, and inspired. Together they have transformed the psyches of tens of thousands of poor farmers and as many acres of farmland in Central America, and they have planted seeds that have grown to a global movement. Together they have enacted healthy community development.

The Human Farm is a tapestry of good development, woven with the biographies and philosophies of a few courageous folk and depicting their human farms—their heads and hands and hearts—the God-given technology of transformation. How this tapestry came to be woven is a story in itself. The first threads were spun back in the fall of 1991, when I worked with Ted Yamamori, president of Food for the Hungry International, and Eric Thor, director of Arizona State University's School of Agribusiness and Environmental Resources (SABER), to develop a symposium on food security and the environment. We hoped to draw attention, ahead of the Earth Summit, to a body of research at the grassroots level: lessons learned by nongovernmental organizations (NGOs) for increasing poor families' ability to feed themselves while respecting the land's need for conservation. The efforts of a dedicated team culminated in a two-day conference at Arizona State University in November 1991 entitled "Growing Our Future: Food Security and the Environment" and attended by more than a hundred conservationists, agriculturists and resource managers from five continents. With financial assistance from Japan International Food for the Hungry, the conference proceedings were published at the time of the Rio gathering by Kumarian Press as a book of essays and case studies.

While preparing the book *Growing Our Future: Food Security and the Environment* with Ted Yamamori and Food for the Hungry colleagues Lisa Leff and Karen Randau, I spent significant time on one essay entitled "The Human Farm: People-Based Approach to Food Production and Conservation," submitted by two Honduran development workers. The writing intrigued me, not because the idea it described—that good development starts with changing people—was brand new, but because the writers' language was fresh, free of jargon, and full of the images and arguments that a poor farmer, a *campesino*, might call his own. For example:

> If the mind of a campesino is a desert, his farm will look like a desert. To overcome perceived conflicts between the objectives of sustainable development and those of environmental conservation in poor countries non-governmental organizations (NGOs) must begin with proper development of the knowledge, motivation and skill of poor people responsible for food production. We must give a "human face" to any food security and conservation program in the world. People are the problem—not the trees, the soil or even the crops—so it is with people we must work to care for trees, manage hillsides, and produce grains . . .
>
> When we take a closer look at the people side of the problem, we see project target populations consist of:
>
> - the economically poorest persons
> - the least educated or trained persons
> - those with the greatest history of failure
> - those who are most disillusioned by politicians, technicians and even development agencies
> - those most attached to traditions
> - those with the greatest number of children
> - those more vulnerable to diseases
>
> By understanding these demographics we can see that the environmental degradation of the campesinos' farms is only another expression of the personal and spiritual degradation of those who work the land. (*Growing Our Future,* pp. 74, 75)

As it turned out, the authors of this essay *were* of campesino stock. Elías Sánchez and Milton Flores, both raised on rural family farms, had become father and son to a movement in Honduras that was empowering campesinos to address their own social and environmental problems.

A light bulb lit up, and I called our editor at Kumarian, Trish Reynolds, to see if she agreed that the story of these men and the movement was a clear snapshot of good development. Trish responded with her first of many votes of support, and I left for Honduras to meet the story's key characters. That trip took place in September and October 1992, a time during which Milton, Elías, and a host of Hondurans and expatriate development workers shared with me their time, thoughts, food, and lodging and allowed me to delve into their lives. You'll meet them all in the story—Miriam and Wilmer, Fernando and Vilma, Sonia, Juanita, Candida, Jorge, Rafael, Francisco, Ramón, Carlos, Margaret, Alfredo, Roland, Loral, Florentino, Roberto, David, Miriam, Aaron, and Mauricio.

I did not set out to describe a spiritual journey, but you will find *The Human Farm* peppered with anecdotes of faith, for good development deeply involves the spirit. Many farmers and development workers I encountered had established new relationships with God and family through better caring for their land. I did not intend to record a philosophy, but you will find a sophisticated folk wisdom in the pages that follow, as personal and community transformation starts in the mind. Indeed, Loma Linda's chief gift to campesinos is new thinking. Nor did I plan to write a manual on organic farming technology, but innovative technologies abound at Loma Linda and have become a basis of experimentation for campesinos as they develop their own farms. They, too, are part of the story.

I did hope to tell a tale of good development. What follows in spirit, philosophy, portrait, and prose is that effort. The lives are the characters' own; the message, their truths transcribed. As for the text, I am weaver, not artist. Where you find beauty, it is theirs; errors, they are my own.

Cast of Characters

The Andrade Family
FERNANDO, peasant farmer from Linaca
VILMA, Fernando's wife
NORA CRISTINA, eldest daughter
SONIA, second daughter
JUANITA, youngest daughter
JILI MICHELE, grandaughter, child of Nora Cristina and her
husband

The Flores Family
MILTON, agronomist from Tegucigalpa, founder of
International Cover Crops Clearinghouse (CIDICCO)
MIRIAM, Milton's wife
ANDREA, eldest daughter
AARON, eldest son
MAURICIO, youngest son

Granja Loma Linda Staff
DON JOSE ELIAS SANCHEZ (or Elías Sánchez), founder and
director (the Spanish title "Don" is honorific)
CANDIDA OSORIO, administrator

JUANA CERRATO, cook and housekeeper
JORGE AMADOR, trainer
JAIME DELGADO, farmhand
LUIS ALONZO MORALES, friend of Elías

Assorted Development Workers

ROLAND BUNCH, former Central American area
representative, World Neighbors; founder of COSECHA
consulting partnership

WILMER DAGEN, cofounder with Elías of Association for
Coordinating Resources in Development (ACORDE);
cofounder of MOPAWI

MIRIAM DAGEN, Central American area representative, World
Neighbors

RAFAEL DIAZ, former supervisor of Elías at Ministry of Natural
Resources; director, World Neighbors/Honduras

MARGARET HARRITT, environmental programs officer, U.S.
Agency for International Development (USAID)/Honduras

ALFREDO LANDAVERDE, former congressman working on
agricultural modernization law

LORAL PATCHEN, U.S. Peace Corps volunteer, Intibucá, Lempira

FRANCISCO SALINAS, agricultural projects director, Catholic
Relief Services, Tegucigalpa

ROBERTO ZEPEDA, extensionist, CIDICCO, El Jute

RAUL ZELAYA, PATRICIA CRUZ, LAURA SUAZO, and
ERNESTO PALACIOS, instructors, Rural Development
Institute, Pan American Agricultural School, Zamorano

CARLOS ZELAYA, administrator, Food and Agriculture
Organization (FAO), Tegucigalpa

Assorted Farmers

JUANITA CERVANTES DE FRANCO, ARMIDA-LARA
ESCALANTE, CAMILO MEJIA, and LUCIO MENJIVAR,
Loma Linda trainees from Lempira

MAURO MENDIZABAL, Pan American Agricultural School
trainee from Bolivia.

JOSE BENITO PONCE and ANDRES PONCE, father and son,
   former Loma Linda trainees from El Jute

RAMÓN ROMERO, philosophy professor and urban gardener

FLORENTINO SANTOS, former Loma Linda trainee from
   Socorro region, Comayagua

GREGORIO and CANDIDA VELÁSQUEZ, former Loma Linda
   trainees from Semane in Intibucá

Others

FRANK ANGEL, New Mexico resident and friend to Elías

LILIANA SANCHEZ, Elías's former wife

SAM ZEMURRAY, president of United Fruit Company; founder
   of Pan American Agricultural School

1

Hunger on the Hillsides

ACROSS THE Texas border and the Rio Grande, past the smokestacks of Monterrey, the pyramids of Teotihuacán, and the sprawling slums of Mexico City, a great highway winds up and down, in and out, through dusty plains and mountain switchbacks, conjuring up the writings of Jack Kerouac, the music of James Taylor, and cravings for ice-cold beer. Somewhere south of the bandits and beauty of Mexico's populous capital, this road creeps over the border to Guatemala, rising to a cold, misty pass so high that clouds sweep through each day, and the locals call it "Alaska." Here, descendants of Mayan Indians remain virtual slaves to tiny plots of land, eking out food for their families and little more. The cheer of

> *For me the saddest thing is a man who works in the fields in May, June, and July, and yet he doesn't have corn or beans. He doesn't have enough to eat in his home. He has to sell everything.*
>
> —DON FERNANDO ANDRADE

Reader's note: Key words and phrases are explained in a *Reader's Reference* at the end of the book.

1

their colorful, traditional garments seems at odds with their doleful stares for strangers passing through. After decades of guerrilla warfare and military oppression in these, their mountains, they have learned not to trust.

And so the highway plunges on between high volcanoes and coffee farms, little villages filled with war widows and heavily armed military posts. Huge diesel trucks, belching smoke and laden with fruits and vegetables, chickens, and lumber, ply this asphalt ribbon day and night. Their drivers, who career along as if hell-bent on an early finish, will almost always tell you that this road has a name, not merely a destination. This is the Pan American Highway, and surer than life and death and free trade it binds together the continents of the Western Hemisphere. One feels somehow strong and full of purpose on it. The gray ribbon passes Guatemala City and veers south again, this time to El Salvador, where it has carried tanks and soldiers to the countryside to slay guerrillas and country folk, rogues and priests, *mano a mano* in a civil war just ended. Transecting the landscape like a meandering stream, the highway with its surroundings is almost too tidy to be war torn. Billions of dollars of U.S. economic aid have come down this asphalt pipeline in the past eight years and rebuilt a tiny nation.[1] But the road is indifferent; it leaves the land of war and plenty as quickly as it came and burrows into a quieter country, Honduras.

Here, in a gentler landscape of hillside farms and banana plantations, it seems that the United States has dropped its calling card, which says, for better and for worse, "We call the shots." Since 1985, the United States has paved this section of highway with more than $1 billion of economic aid.[2] It gained contra forces a safe haven during Nicaragua's civil war and has greased the wheels of enterprise for transnational companies. In private, a Honduran will tell you that his nation's first president is not the fellow who shows up for Central American summits but rather the U.S. ambassador. The second president is the commander of the armed forces. Civilian presidents rank third. But they are all important travelers on the Pan American Highway as it winds south of the nation's hilly capital, Tegucigalpa, and on to the next country, and the next, to the heart of Panama.

Still further, an hour east of Tegucigalpa, the road forks. To the left, the highway curves up through a mountain pass to a prestigious institution, the 15,000-acre Pan American Agricultural School, which turns out the cream of Latin America's agronomy crop. Nurtured on Western technology and boot-camp discipline, its students have a mission: to improve the region's agricultural production and exports. To the right, a route less often traveled tumbles down a stony path to a village called Linaca, where unschooled farmers till steeply sloped family plots of less than three acres. Their mission, like that of the world's two billion village farmers, is simply to feed their families. One Linaca farmer, Don Fernando Andrade, has been trying to do so for most of his sixty-four years.

At a glance, there is not much to see in Linaca. Its tiny town square is wet from autumn rains, empty of life, and strewn with garbage. A policeman slouches in the doorway of an adobe cottage off the square and eats an ear of roasted corn. Chickens cluck and scavenge in a neighboring yard full of mud and refuse, while pigs wallow in a smelly sty nearby. No, there is not much to see in Linaca, except Don Fernando's corn. It is harvest time, and Don Fernando's stalks, drying in the field, stand three feet taller than his neighbor's. Its ears are double the size of the small cobs that typically result from Linaca's poor soil and dry spells. If you ask Don Fernando how he has managed this, he will grin and point not to improved technologies, of which his farm has many, but to his heart, his head, and his hands, which he calls his "human farm." And if you sit a while, he will tell you how it wasn't always this way.

The Andrade family plot, like two-thirds of Honduran farms, is located on a hillside with a more than twenty-degree pitch and is covered with poor, clay-ridden soil. Like other small farmers, Don Fernando grew up working hard to grow corn and beans on his tiny acreage, sowing and harvesting by hand a much greater percentage of his land than did the bigger farmers on better, flatter ground with hired labor and machinery. And like other small farmers, Don Fernando was rewarded for his efforts with lower production and loss of much harvested grain to rats. After feeding his family, there was little left over to exchange for other nutritional, household, and farm needs.[3]

From dismal land conditions like these rise the tragic statistics of Honduras's rural sector. It accounts for 2.8 million of the country's 4.4 million inhabitants, and the infant mortality rate reaches 13 percent in some regions.[4] UNICEF's 1990 situation report on children in Honduras found that 93 percent of rural households were living on less than $24 a month; throughout the countryside, nearly half the children under age five were malnourished.[5] Imagine working harder, longer, and hungrier hours, days, and weeks than your neighbor and finding that you have less to show for it, sick children, little income, and no idea how things could be different. If you can imagine that, you can imagine the plight of the Central American village farmer, the campesino, the Andrades.

Don Fernando, his parents' oldest child, went to school until the second grade. Then his father pulled him out to work in the fields. "That was the tradition in our pueblo, no more than two years of school for the son," says Don Fernando. "Then the son must work for his father." And the work was backbreaking: first burning the fields to clear them, then plowing steep slopes either with oxen or by hand; next poking a hole in the ground for each corn or bean seed; weeding; watering by hand in dry spells; and finally harvesting his meager production. For years, Don Fernando helped his father and lived with his family in a cottage above the cornfields, an hour's walk up an age-old, carved stone path from Linaca. For years he toiled, but the disappointing results never quite rid his family of the specter of hunger. Then, at the ripe age of thirty-three, a blessed distraction came to Fernando. Her name was Maria Sabas Zelaya.

Maria Sabas, or Vilma as her friends call her, was a local beauty with an infectious laugh and countless admirers. She grew up on a small farm down the valley from Linaca, halfway to the Andrade's hillside plot. She was Fernando's neighbor, and yet, as custom and coincidence would have it, the two had never formally met. That changed one evening shortly after Vilma's nineteenth birthday when she attended a Linaca fiesta at the home of Fernando's cousin. Vilma walked through the door, and Fernando looked up and saw her. "Would you like to dance?" he asked. He danced with several young women that night, but it was on Vilma's household that he called the next day. He called again and again and again. Within the year they were married. By the next year, they had their first child.

As of 1980, rural Honduran families were having an average of eight children, or double the birthrate of families in urban areas.[6] The rural family's strategy is to have enough children to ensure that at least a few creep by the statistics on malnutrition and infant mortality to reach adulthood so that they can work the land and take care of their parents in old age. Children are the social security of the poor. Many studies have shown that as a population's income and education increase, and thus its ability to combat disease and malnutrition, its birthrate decreases. In fact, there is a significant correlation between increasing literacy, especially women's, and lower population growth.[7] The best answer to overpopulation and poverty lies in educating women and helping to strengthen revenues in poor communities.

Given these rural trends, Don Fernando and Vilma took an unusual and prescient step: They decided to have children according to their economic means rather than their needs. As far as Fernando was concerned, that meant one child only—until the child turned out be a girl, Nora Cristina. Hoping for a boy to help on the farm, the Andrades tried again seven years later and had another girl, Sonia Natalia; finally, seven years later, they had another, Juanita Maria. As did most women on the hillside farms of Honduras, Vilma delivered her girls alone, in her bed, after she had seen to it that the chickens were fed, the washing done, and enough tortillas made to see her family through her hours of labor.

Fernando loved his daughters but longed for boys to help him work the farm. The land seemed to be producing less and less the more he worked. Frustrated by his plight and with nothing to cling to but tradition, Fernando forbade his girls to go to school, took heart in carrying a well-hung pistol, and resigned himself, like so many others with poor soil and few job skills, to the likelihood that he would one day give up his farm and move on, family in tow, in search of better land to slash and burn, plow and cultivate.

By the early 1970s, the Andrades had moved from the home of Don Fernando's father to a little adobe cottage in Linaca; his brother and sister-in-law moved in across the road. Don Fernando planted corn and bean fields around his new house and continued the daily trek to his family's hillside plots to farm with his father. With an ox-drawn sled, which Fernando had carved from a single tree trunk, the two men would drag large rocks and stones from the fields. With an Egyptian plow made from a metal plate and a tree branch, they would furrow rows for planting. But each year, as they removed the stones and burned the weeds to prepare the land, the rain would wash more and more topsoil down the steep hills and into a brook below. Less and less corn would grow.

One day, the hill played an even crueler trick on Fernando's father as he labored for his unwilling crop. The old man tripped on a root, slipped in the moist clay, tumbled down the slope, and crashed on his back into a rock that he had not yet dragged from the field. Fernando carried his father from the field back to Linaca. He transported him all the way to Tegucigalpa to the hospital, but the old man never healed; nor did the land repent. Fernando's father died, and his family had less food than ever.

"This is a very sad house," thought Vilma. "There is no one to help us, and we have no new ideas to make life better." Although they had corn and beans from the field, the Andrades had little to exchange for supplementary foods, meats, or dairy products, and the children's health suffered. Fernando might have succumbed to the temptation, as do many hillside farmers, to sell all his food for cash, mortgage his next crop to moneylenders, and so enter a vicious circle of poverty and indebtedness, but his father had taught him better. All Fernando knew, he had learned from his father: to burn the weeds, to sow the seeds, to keep the food you needed to survive. All this he did, and it was not enough. His prospects of survival were washing away with his topsoil, and his father hadn't taught him the cure. And so, around 1978, Fernando prepared to leave Linaca and his roots and seek a new home on virgin land. This was his only hope, he thought. In fact, he was certain of it, until another blessed distraction came into his life. His name was Don José Elías Sánchez.

Notes

1. From 1985–89, the U.S. government provided El Salvador a net $2.05 billion in grants and credits, making it the third largest bilateral recipient of U.S. economic aid for that period after Egypt and Israel. U.S. Bureau of the Census, "Foreign Commerce and Aid," in *Statistical Abstract of the United States: 1991* (Washington, D.C.: Dept. of Commerce, 1991), pp. 799–801.
2. From 1985–89, the U.S. government provided Honduras a net $1.01 billion in grants and credits, making it the sixth largest bilateral recipient of U.S. economic aid for that period. Critics have called the assistance a U.S. *quid pro quo* for Honduras's willingness to serve as the jumping-off point for contra operations in Nicaragua. U.S. Bureau of the Census, *Statistical Abstract: 1991,* pp. 799–801. Interview with Vincent Cusumano, U.S. Agency for International Development Director for Rural Development Office, Honduras, Sept. 30, 1992.
3. G. Galvez, M. Colindres, T. M. Gonzalez, and J. C. Castaldi, "Honduras: Caracterización de los productores de granos básicos," *Temas de seguridad alimentaria no. 7* (Panama City: CADESCA, 1990), pp. 20–38.
4. UNICEF, *Children in Honduras* (Guatemala City: Dixon Print, 1990), pp. 36–37. This is the most recent international report that details Honduran child health statistics by region.
5. UNICEF, *Children in Honduras,* pp. 43–48.

6. UNICEF, *Children in Honduras,* pp. 5–6.
7. Statistical regression exercise run by Professor Anil Gaba, European Institute of Business Administration (INSEAD), Fontainebleau, France, Feb. 1993.

2

A Farm Called Loma Linda

> *Human misery is not lack of money, it's not knowing who you are . . . Dissatisfaction is the beginning of change.*
>
> —DON JOSE ELIAS SANCHEZ

DON JOSE ELIAS SANCHEZ is not a saint. He is a man like any other, with feet of clay. And he is not above using them to give swift kicks to those who cross his purposes. Friends and foes alike bear his bruises, but many agree that he has a saving grace: He has done more to improve the lot of campesinos in Honduras, perhaps in all of Central America, than any living soul. And he was soon to turn his brand of harsh love on Don Fernando.

This love, which Don Elías uses to elicit change from farmers and to exact it of his peers, is born of his own struggle out of traditional rural farming to higher learning and "human farming," a term he coined to describe innovative, loving, hard work to cultivate living soil. Born in 1927 on a 2,500-acre farm in Honduras's southernmost department of Choluteca, Elías never knew the "privilege" of being yanked from second grade to work the fields. His father, a dominant personality, knew good labor when he saw

9

it and forbade school altogether, despite his wife's hopes to educate their children. "If you want to make money, don't go to school," said father Sánchez. "But I want to learn," thought Elías. Through a hard twist of fate, he got his wish. His father became ill and passed away at the age of forty-five, and cruel custom deprived his mother of the land they had farmed. Her husband's brother took the acreage, and Señora Sánchez, a tiny woman with great resolve, moved her five children to town, where, though poor, they could attend school. In Oroquina, at age fourteen, Elías entered the first grade. That single accomplishment most surely changed his life.

The boy was hungry for studies, frustrated by the time he'd lost, and in good company. About a dozen older farm boys entered first grade with him. Elías, with his brothers and sisters, walked nearly four miles each day to and from the schoolhouse, and he thrived. In 1946, having climbed through ten grades of schooling in six years, Elías won his first of many scholarships. The Honduran government named him to study in a unique postsecondary teachers college, called a rural normal school. These schools had been established on the advice of the U.S. government as the Truman administration strove, after World War II, to create a new world order that would strengthen allies at their grassroots and entrench American-style democracy. With a regime of strict, energetic teaching, much of it by imported Puerto Rican instructors, the rural normal schools fostered an appetite for learning among country children that propelled many of them on to graduate studies.

Elías's journey down this road began atop a truck full of chickens, pigs, and grain that carried him and his bundle of possessions, jouncing, from Oroquina to the Varones Normal School in Comayagua. At the college, Elías found that in addition to academics he was required to practice sports, cooking, cleaning, washing, carpentry, blacksmithing, and his former nemesis: farming. Ironically, after graduating three years later, it was the latter course that helped him make an impact as a rural elementary school director back in Choluteca.

"What am I going to do?" Elías asked himself, as he arrived at his first teaching post, El Corpus, to find a classroom of eighty kids from four grades in a barnlike structure with long wooden benches. "This environment is good for bats and owls," he surmised, "but not for children." And so Elías set to work to change it. He chopped up the benches into smaller seats and opened up windows in the walls of the school. Best of all, thought the students and their parents, he began a school garden. The plot afforded the children hands-on, practical lessons in science, nature, and teamwork, and their achievements were tangible both on the ground and in their diets. The garden became the talk of the town, then the region. "You are different, Don Elías," the parents told the farm boy turned teacher. "Our children hated to sit in school, but now they hate to come home." Most seemed to agree that the five-foot-five-inch dynamo had truly earned his salary of $22.50 a month.

If Elías had been unsure of himself when he first arrived in El Corpus, he was firmly rooted and enjoying the fruits of teaching three years later when he was transferred in 1952 to the northeastern Choluteca town of Duyure, with a monthly salary raise of $7.50. By now, Elías had learned an important lesson: Parents were the key to teaching children; indeed, teachers could only supplement what parents were teaching at home. And parents tended to become involved in a school not through math but through activities: sports, music, or even school gardening. Moreover, as parents, who were often illiterate themselves, became involved with the school, they too received an education. The process demystified higher learning and increased community enthusiasm for formal studies. "Build a garden," Elías would assert, "then you can get the support of parents; then you can teach."

But parents weren't the only ones observing and approving Don Elías's creative approach to teaching. After a year teaching in Duyure and a year working for a government agency in agricultural extension, a telegram from the Honduras Ministry of Education informed Elías that he had been selected to attend an international teachers college, the InterAmerican Rural Normal School in Venezuela, with a scholarship from the Organization of American States. The purpose of this school, again a product of the Truman era, was to create teams of professionals in health, homemaking, technology, agriculture, and administration to bring a practical, holistic teaching method to Latin American schools.

Elías was to study agriculture, and he did, but a communications course most profoundly advanced his thinking. "Don't try to work with a cow," chided his linguistics professor, Aristobulo Pardo, "work with the people who milk the cow!" Elías saw that communication was a cycle between teacher and student, highly dependent on what the listener brought to learning. And he discovered that he was lacking in knowledge as a listener: He had never read the classics, whose themes were bandied about by fellow students from Chile and Argentina. Discovering the void in his education, the boy who started first grade at fourteen began quietly noting the titles of literary classics, procuring them, and reading them in his precious free time. In them, he discovered messages that struck at the very essence of humankind's purpose and promise. He learned from Mark Twain how America was settled; from Tolstoy, how nations evolved. Meanwhile, other instructors were drilling the students after classes in newfangled sports such as basketball. Between his course work, his classics, and compulsory sports, Elías crawled into bed each night happily exhausted.

In 1956, the year a twenty-seven-year-old black preacher named Martin Luther King, Jr., grew to fame as leader of the American Civil Rights movement, Elías returned to Honduras to apply his learning to liberating rural teachers from a top-down, hierarchical classroom tradition. The government assigned him to the women's normal school of Villa Ahumada in the department of El Paraíso, northeast of Choluteca. He had two missions: to train future teachers in the ways of agriculture and animal husbandry, and to see to it that the school maintained its vegetable production to supplement students' diets. When Elías arrived, the school was purchasing most of its food; by the time he left, it was producing it.

The transformation took place not through technology but through applying the lesson of his communications professor in Venezuela—through focusing on the students, who were being treated merely as a free labor force. "Let's give students some incentives," said Elías. "Let's assign a percentage of the school's production profits to them." The students, treated by the school's twenty teachers as valued coworkers instead of serfs, pitched in on the little school farm and earned pin money to buy their class rings and other personal effects. The school prospered, as did its students. Their rapport with teachers evolved from blind acquiescence to engaged learning. The institute's hierarchy had been turned on its head: Elías held up the janitor as a role model for leadership and bade teachers to emulate his servanthood. Once again, more than just local eyes blinked in interest: The United Nations Educational, Scientific, and Cultural Organization (UNESCO) got wind of Villa Ahumada's success and held the school up as an international example of effective teacher training.

As for Elías, he was about to benefit, once again, from American interest in his region. Honduras's Service for Inter-American Cooperative Education (SCIDE) was funded by the Inter-American Cooperative Administration, a Truman-era precursor to President Kennedy's Agency for International Development. SCIDE hired Elías to supervise a national training program for elementary school teachers. In conjunction with the SCIDE initiative in education, the Inter-American Cooperative Administration funded initiatives in primary health care and agricultural development. As Don Elías is fond of pointing out, it was a time when the U.S. government seemed concerned with fostering truly good development in the Americas. Even the agency's name implied an equal partnership between the United States and its neighbors. "It saw development like this," Don Elías would say, clasping together two soil-covered hands.

In Elías's case, the U.S. helping hand lifted him to heights of learning that his mother had never dared hope for him. Elías grew and developed as a leader in education, attending workshops with his SCIDE coworkers and then replicating the lessons learned for teachers throughout the country. Elías became convinced of good development's multiplier effect and eventually was promoted to assistant to SCIDE's director, an American named Grace Scott. Within two years, he was awarded another scholarship, this one by the newly formed U.S. Agency for International Development

(USAID), to study for a bachelor's degree in Education at the University of New Mexico in Albuquerque.

Speaking only rudimentary English on his arrival in the United States, Elías survived and got his B.Ed. by being what he has become known for: consistent, persistent, and insistent—in other words, pushy. The New Mexico years were heady ones for Elías. He broadened his intellectual and political horizons, even as America's seemed to be shrinking. In his first years there, the United States suffered a humiliating defeat at the Bay of Pigs in an attempt to overthrow Communist Cuba, and the Soviets won the space race by sending *Sputnik* into orbit. Chagrined U.S. scientists and politicians scrambled to regain the lead in technology, the initiative in foreign affairs. Almost overnight, the United States transformed the national science education curriculum in the name of national security and forced ripples of change as far away as the University of New Mexico, which switched from the decimal system to binary mathematics. Though confused, Elías and his expatriate peers evolved with the curriculum. Toward the end of his studies, Elías, like most in America at the time, was sickened by the news that a lone gunman had assassinated President Kennedy in Dallas. Those from countries like Honduras, which were still tumbling in and out of civilian rule and seeking the meaning of democracy, were all the more shattered by the discovery that bloody assassins could dictate political fates in the United States.

Elías came back to SCIDE in 1964 a thirty-seven-year-old bachelor. He returned to Albuquerque for a master's degree four years later with a young wife, Liliana, and an infant son, José Elías. He had waited a long time, eyed the field, and wed a social worker. His new responsibilities as husband and parent were made easier by continued sponsorship from the U.S. government: USAID granted him another scholarship to the University of New Mexico, and SCIDE kept him on salary. The boy who had almost missed out on school altogether never had it so good. He made himself a promise: No matter what opportunities lay ahead in America, he would bring his training back to Honduras. "Whatever happens," he said, "I will come home."

On his return to New Mexico, the Honduran government appointed Elías its honorary consul in Albuquerque, a move that opened the doors of upper-class America to the diminutive Latino. The year was 1968: war raged in Vietnam, and U.S. minority groups, still fighting to end discriminatory laws and practices, lost to gunmen two of their most powerful champions: Martin Luther King, Jr., and President Kennedy's brother, former U.S. Attorney General Robert Kennedy. The assassinations marked a year of prolonged urban violence and social upheaval. Throughout, one American minority group member, Frank Angel, became the lens through which Elías viewed not only the social fabric of America but also the misery and indifference in his own country.

One of the first Chicanos in New Mexico to receive a college degree, Frank, who had consulted on a project in Honduras, reached out and took in Elías and later his family during their years in the Southwest. Frank taught Elías American specialties from modern home-building to Chicano cuisine; he took him to local board of education meetings for on-the-job training in school administration. Elías became increasingly aware of the rare set of skills he was acquiring and redoubled his commitment to repatriate them. After graduation in 1969, with sociology, anthropology, education, and philosophy courses under his belt and a master's degree in educational administration, Elías sought a speedy return to Honduras. Civil war in El Salvador blocked the overland route, so the Sánchez family threw their luggage atop their car and tore down the highway to New Orleans to catch the next ship home. In his eagerness, Elías took a wrong turn in Texas and, three hours later, discovered that he was headed back to New Mexico. The man who would later note that Hondurans who studied abroad often came home so pompous that they "confused a parrot with a chicken" had a hard time telling East from West in his excitement to repatriate.

In the end, Elías's mother back in Choluteca was as strong a magnet pulling him home as his vocation. Just four and a half feet tall, she was made, Elías would tell friends, "of hard wood." When his father died, he had watched his mother beg for credit at the bank, only to be refused. He had seen the strong woman break down and cry, and it tore his heart. "When I get some money," he promised himself, "I'm going to give it to my Mama." And indeed, he became his mother's adviser and supporter until her death. But there was one promise to his mother that Elías chose not to keep: that he refrain from marriage until after she passed away. "The problem is, we have Indian blood," Elías would later say, "and we never die." His mother lived to be almost a hundred, and never forgot, nor did her son, that she was due the first fruits of his love. It was a reality with which Elías's wife, Liliana, was forced to contend.[1]

On his final return to Honduras, Elías was appointed director of the National School of Agriculture, again through SCIDE, but SCIDE was now on the wane. A year later, the U.S. government agency was phased out, and with it plum salaries and company vehicles. Elías and his colleagues found themselves pariahs in the government job market, supposed moles for the U.S. Central Intelligence Agency. "We are in hell," thought Elías. "The [Honduran] government won't give us a chance." Elías met this dismal realization with the wisdom of his old school gardens: "A gardener can put his plant where he wants. So if one day God tells me, 'Don't plant yourself here,' I must plant myself in another place." Thus Elías, beneficiary of a lion's share of international opportunities, took a new tack and dedicated his energies to national institutions. He worked first at the National Autonomous University of Honduras, where any taint of his U.S. agency employ ostensibly diminished, then joined the Ministry of Natural Resources as a regional program coordinator in rural development. More importantly, he began to think about how he could use a patch of hillside wilderness he had purchased, just half an hour from the capital, to teach hillside farmers how to use their heads and hearts and hands to improve their lives.

One of his first steps toward this end was to tap into a nascent movement of grassroots Honduran development organizations, largely outgrowths of church ministries to assist the poor. By 1972, the Catholic Church had launched a village-based Bible study movement, Delegates of the Word, which had a social outreach component that, among other things, advocated that farmers cease burning their fields and actively conserve their topsoil. The Protestant churches, largely evangelical, had formed a nonprofit organization called Diaconia, derived from the Greek for "service" (and from which we get the English word "deacon"), connoting help for physical and spiritual needs. U.S. troops in Vietnam were in their final months of fighting, and assisting Diaconia's cause was a Mennonite conscientious objector from Alabama, Wilmer Dagen, who chose development work over the draft and got hooked. He hired Don Elías to help Diaconia in Honduras's northern regional capital of San Pedro Sula. Don Elías combined government work in rural extension with running training sessions on behalf of Diaconia. In fact, he wore private- and public-sector hats in a way that is typical of many developing country civil servants, whose low government wages often necessitate both moonlighting and family farming to make ends meet.

Elías called himself the "brown Gringo" in light of his U.S. education, and Wilmer was a "white Latino" who saw his role as aide-de-campe of Honduran staff.[2] As a duo they managed to relate well to national and international donor and government officials, city folk, and rural farmers. Together they visited rural communities throughout the country and trained farmers in the ways of soil conservation and self-respect. Wilmer's American wife, Miriam, had been in the country a matter of days when Elías showed up on her doorstep at 2 A.M. to take Wilmer on a field trip. She'd come to Honduras a starry-eyed bride, only to find that her competition was a short, feisty Hondureño. "My rival is Don Elías!" she laughed to herself. Unable at times to compete with Wilmer's zealous friend, Miriam began working three jobs and lunching with Elías's wife, Liliana, now the mother of two.

In 1974, Elías and Wilmer cofounded another nonprofit organization, the Association for Coordinating Resources in Development (ACORDE), and Wilmer became Elías's assistant and codirector. But from the beginning, the agency, which organized training sessions for campesinos in human development and agricultural improvement, was virtually a one-man show, dependent on

Don Elías's unique and strong personality. With his short temper and stubborn resolve firmly in evidence, Hondurans who joined his cause didn't stay around for long. Lacking full-time staff, ACORDE's reason for being was simply to promote, coordinate, and channel national and international resources to Honduras's rural sector. Some of the first resources on which Don Elías set his sights were homegrown, those of the Ministry of Natural Resources. That same year, the government had made him director of its national training program for agricultural extension workers, and his self-appointed mission was to increase their credibility among farmers. "Professionals in Latin America lack commitment," he observed. "Our big task right now is not to improve their handiwork, but their hearts."

His supervisor at the ministry, Rafael Diaz, who sympathized with Elías, called him a "'voice crying in the wilderness" for a grassroots approach to rural extension. However, over several years, the scope of Elías's job allowed him to influence the wilderness in which he cried, putting his stamp on about 600 civil servants and 3,000 agriculturists. Seeking practical opportunities to train his charges and make contacts for ACORDE, Elías would sniff out rural communities ripe for training sessions, then bring his extensionists to the campesinos. When farmers responded positively, he organized field trips to show them pioneering techniques in hillside farming. A frequent destination was the training turf of Guatemalan Marcos Orozcos, a staff member of the nonprofit international development organization World Neighbors. Working in and for his country, Marcos taught farmers to conserve topsoil by planting in terraces, digging contour ditches to catch and channel rainwater, and creating live barriers of grasses, bushes, and trees to hold the soil. No doubt, many Honduran farmers drove with Elías the day's journey to Guatemala only for the opportunity to see some unfamiliar countryside. But they left having seen men and women, not unlike themselves, whose ideas and efforts had tamed steep hillsides and transformed farmers' self-confidence.

And so it was that one spring day in 1978, Don Elías followed his nose to the little village of Linaca and the whitewashed adobe home of Don Fernando Andrade, a man of about Elías's age who had all but given up hope of improving his lot. Hondurans have an earthy expression for detecting a man's true substance: "From a distance you smell, but up close you stink." To Elías's olfactory glands, Don Fernando positively reeked of potential. Husky and introverted, Don Fernando would not even look Elías in the face during their early conversations. Instead, he cast his eyes on his dirt floor, his hand automatically on his holster. He had little experience with government officials and no reason to trust them. "He has more belief in his pistol than in his intelligence," thought Don Elías, but he knew that that would change. "Señor," he said, "I am Elías Sánchez, and I would like to work with you to grow better corn."

Although suspicious, Fernando was impressed that Don Elías, a man with social and professional standing, was sitting in his home and talking to him as an equal. He had been on the verge of leaving his village to look for better soil, but he liked the idea of holding a farmer training course in Linaca and agreed to spread the word. About half a dozen farmers, including Fernando's brother, Edilberto, turned out for the course, and they were surprised to discover extension workers teaching not modern technology but practical, traditional techniques for soil conservation. They learned terracing, a technique that had been widely used hundreds of years earlier by Andean farmers, the Incas. They learned to position grasses, trees, and hedgerows to hold the terraces, to dig contour ditches, and to minimize their labor by preparing only a small alley of soil on each terrace for planting—a technique that Don Elías himself had developed, which he called "in-row tillage."

The farmers understood that the soil was the food their crops needed, just as crops were the food their children needed. But they did not understand what made the soil healthy nor, for that matter, what made their children unhealthy. They knew that the big farmers used chemical fertilizer and pesticides, but most small farmers could not afford them. They were surprised to learn that their soil also needed food, and that its favorite food, organic matter, lay in the streets and yards of garbage-strewn Linaca. Don Elías's course spent an entire day teaching farmers to make organic compost piles from old papers and other trash, corn stalks, vegetable peelings, and bean plant residues, and then showed the farmers how compost could be mixed with soil in the narrow alleys of in-row tillage to create a fertile strip for planting corn and bean seed. Don Fernando was highly impressed. At last he and Vilma saw a flicker of hope for surviving on Linaca's poor soil. This was an idea that they could afford to try. The course showed him that although his father's technique of burning fields to prepare them for planting had helped kill weeds, it had also destroyed the very organisms and residues that kept the soil alive—the old stalks and dead leaves, the maggots and worms that broke them down into rich organic food for the corn seed. Don Fernando issued an edict to his little family: From then on, their yard should be clear of litter; all trash was to go into his new organic compost.

For Fernando, the soil conservation methods would mean more tedious hand labor since the oxen were too large to plow narrow terraces. But it also meant higher yields on less land. If all went well, he could cultivate the land around his Linaca home and reap a greater harvest than he had by sowing both his own field and the steep slopes an hour away that his father had farmed for years. Elías's nose had not led him astray. Don Fernando stank all right. He stank of love for his land, of a willingness to work hard with his hands, and of a mind yearning to wrap itself around new ideas. He stank of such promise that nothing in Linaca was going to stop him from growing the tallest corn in the village. At the end of the course, Don Fernando and Vilma invited the other farmers to their home for a closing ceremony. And when Elías bade him farewell, Don Fernando looked him in the eye.

Elías took a deep personal interest in Don Fernando's progress and visited him regularly. He took him and other farmers to see Marcos Orozcos's work in Guatemala. As Elías and Fernando talked and observed changes on the Andrade farm, Elías's plans for his own piece of land took shape. Could farmers survive on these hills? Yes! Did Honduran professionals teach hillside farming? No. Elías, the iconoclast, would organize a training center for small farmers on land just as hilly and daunting as their own. There, they could come, work, learn, and see that using merely their human farm machinery—heads, hands, and hearts—they could lure rich crops out of the stony hillsides. They would draw on positive traditions of the region as old as the irrigation systems of the Maya. Farmers would see that the machinery of their human farm held the key to solving other problems—their children's health, their housing, their happiness. He told Liliana his plans. Did Elías see, she wondered, that his zeal for such a farm would bring little happiness to his own family?

But he was convinced of his cause. Just as Elías had found that school gardens were the secret to getting parents involved in their children's education, so this training farm could be the secret to drawing rural farmers into educating their families and communities. It was an ambitious thought, and it required more resources. Through ACORDE, Elías arranged to bring Marcos Orozcos's organization, World Neighbors, into Honduras to work in grassroots rural extension. And with his own two hands, a big heart, and a small loan from World Neighbors, he began to organize Granja Loma Linda, the Beautiful Hillside Farm, because he saw the soil and knew that it was good.

Notes

1. Interviews with Elías Sánchez and Miriam Dagen, Sept. 24– Oct. 4, 1992.
2. Interview with Wilmer Dagen, Sept. 30, 1992.

3

Of Corn and Cover Crops

WHILE ELIAS Sanchez cruised the Pan American Highway in the 1970s, sniffing out towns ripe for change, another man, still in his teens, sped down the highway past Linaca and over the next mountain pass to the sprawling, neatly groomed properties of the Pan American Agricultural School. Attending the famous institution, more commonly known by its former hacienda name—Zamorano—had been a boyhood dream of young Milton Flores.

Coming out of Zamorano we had three ideas: to own our own farm and make a lot of money; to go get a master's and become a scientist; or work for the government and get a paycheck.

MILTON FLORES

Milton, like many young Latinos, knew that a Zamorano diploma usually meant a one-way ticket to a good job, an academic career, or, best of all, one's own commercial farm. Indeed, it was for the future of bright young men like Milton that another dreamer, American banana magnate Sam Zemurray, first built the institution.[1]

A Russian immigrant who came to the United States as an eleven-year-old in the late 1880s, Sam Zemurray was fascinated by tropical fruits, which were foreign to Eastern Europe. He earned a fortune by the time he was twenty-one selling imported bananas out of Mobile, Alabama. He then went on to settle in Puerto Cortés on the northwest coast of Honduras, establishing the Cuyamel Company, which was bought out in 1930 by his large competitor, the former United Fruit Company (UFC). Next, he quietly acquired shares in UFC, until one day he announced to the board his majority ownership and became president. The United Fruit Company (today Chiquita Brands International of Cincinnati, Ohio) changed the face of Central American agriculture, acquiring huge tracts of the region's best farmland for its plantations. It worked closely with national governments, many of them military, receiving tax breaks and railroad concessions. It both created employment and gained profit margin on cheap labor, in keeping with U.S. turn-of-the-century capitalists such as Stanford and Vanderbilt. Not all politicking was overt: In Guatemala, the UFC quietly backed a coup in the 1950s to oust the country's first democratically elected president, Jacobo Arbenz, whose land-reform policy included confiscation of lands from the fruit magnate.[2]

Nevertheless, like Stanford and Vanderbilt, UFC's Zemurray decided to give back to the region that had lined his coffers and that he called home by founding an institute of higher learning. In 1942, Zemurray founded the Pan American Agricultural School on a 2,000-acre farm purchased by UFC from the wife of Honduras's then-strongman, General Tiburcio Carias Andino.[3] Its mission was to improve Central American agricultural development through the training of good, practical agronomists: its mottoes were *aprender haciendo* (learn by doing) and *labor omnia vincit* (work conquers all). The result was an agricultural boot camp, where students rose at 5:30 A.M. to work the farm and went to bed, exhausted, at 10 P.M. Sandwiched in between were afternoon classes in "modern farming" and obligatory study hall.

Until 1958, Zamorano was run by Americans and financed entirely by UFC,[4] whose company policy forbade the hiring of graduates to ensure that the school's prodigies sprinkled themselves throughout Central American agricultural enterprise. In the early years, Zamorano staff scouted students on horseback, recruiting from farms and villages: no high school diploma was required. Dorm floors were made of brick and clay, electricity took the form of naked lightbulbs, and students cleared land and built fences to develop the farm. Graduation for these early students was far from guaranteed: In 1947, for example, only half the entering class of sixty students survived the three-year agronomy program to graduation.[5] Those who did prevail went on to become influential businessmen, academics, and government ministers.

After 1979, under the direction of a dynamic Zamorano graduate, Dr. Simón E. Malo, the school underwent a dramatic expansion and modernization to 15,000 acres and 115 courses, including studies in rural development and appropriate technology. But when Milton Flores entered the school in 1974, he joined a class of just sixty-three students, one-quarter of them Hondurans, who believed that the future of Central American agriculture lay with John Deere tractors and the mysteries of North American chemical farming.

Like Don José Elías Sánchez, Milton came from campesino stock. His grandfather farmed in a small village near the 1950s mining center of Minas de Oro, in Honduras's central department of Comayagua. Three of Milton's uncles became farmers in the region, but Milton's own father, unlike Elías's, had urban designs for his children's education. He married and moved to the capital, where he found work as a mechanic; his wife worked as a nurse. But just as a detour in the plans of Elías's father brought a school-hungry fourteen-year-old out of the family fields, a detour in those of Milton's father sent a quick-witted child back to the family farm. In 1960, when Milton was just five years old, his father was hospitalized with a concussion after an accident at work, and Milton was sent back to Minas de Oro to live with an aunt.

His father's initial months of treatment stretched to six years of medical operations, during which Milton grew up a rural farm boy, milking cows and planting and harvesting corn and beans. Although he never went hungry, there were times when his relatives had little to eat but beans and tortillas. When at age eleven Milton returned to the capital to live with his parents and attend a private secondary school, he knew what he wanted more than anything: to be an agriculturist, an *agronomo de Zamorano*.

Milton was a gifted student and passed Zamorano's entrance requirements, but paying the school's steep tuition (in 1992, the equivalent of $6,280 a year) required external resources. His parents took out a loan, and Milton's father, proud of his son's bright future, accompanied him up the Pan American Highway, down a majestic drive lined with giant Benjamin fig trees, and into Zamorano's main quadrangle of whitewashed brick buildings. "Take care son," he said, and slipped Milton some spending money for his first months in boarding. Milton looked in his pocket and smiled. His father had given him the equivalent of $2.

It didn't take long for Milton to seize the essence of Zamorano: disciplined study, hard work, and obligation to excellence. Distractions were few: The students wore blue jeans and blue work shirts, and the only available female companionship lurked, for daring initiates, in the bamboo grove near the sawmill on Saturday nights.[6] Milton's morning duties on the farm ranged from carrying stones out of the fields to planting grains, spraying the citrus groves, fertilizing vegetable plots, harvesting coffee, feeding cattle, milking cows, butchering meat, and making cheeses. For students who wearied under the routine, there was a quaint reminder of the relative leisure afforded the modern farmer by mechanization: The school's first tractor, a John Deere, was immortalized on a brick pedestal near the campus gas station with the inscription, "Rest has triumphed." Honduran students like Milton got so caught up with the thrills of mechanization that they overlooked its limits— tractors couldn't function on the twenty-degree pitch slopes that covered most of their country! The students' obsession became *increased production by any means*. Much like American agricultural schools in the mid-twentieth century, Zamorano fought to produce more crops in greater concentration without much thought to ecological integrity. "We can grow more bushels of corn in one field than I ever dreamed possible," thought Milton. It was just a question of fertilizer and machinery.

Zamorano was a tough row to hoe for Milton—academically,

physically, and emotionally. But as his three years drew to an end, he was proud of his accomplishments. Shortly after his arrival, the German government had offered him a scholarship, alleviating his parents' financial burden, and throughout the program he had passed each academic hurdle with success. Now, with a Zamorano degree in hand, another dream leapt within his reach: Three days after his graduation, held under the arcade of Benjamin figs through which he and his father had first entered Zamorano, Milton began working for the U.S. Agency for International Development's (USAID's) National Cadastre Program. "My main goal," he would later confide, "was to . . . have a nice car!" At age twenty-one, he was put in charge of eleven other professionals, all working in the department of natural resource's soil section on a project that took inventory of the country's natural resources to make maps for future regional development initiatives. Thus, Milton became involved in development work. Yet as far as he was concerned, his achievements were purely technical. On his first field trip, USAID gave Milton a $118 advance and, after his first month of work, a $300 paycheck. He was now taking home, he realized with awe, much more than his father had earned for a great part of his life. Two years later, he moved to the Honduran Ministry of Natural Resources. He was proud of the task he was doing, and he liked the fact that work for the government, then under military rule, was low on politics. Milton had arrived, it seemed. Yet somewhere, tucked in the back of his mind, lay a niggling doubt that he had not really arrived, that he might, in fact, be lost.

"What is it?" thought Milton. "Why am I so depressed?" He was at Tegucigalpa's International Airport, watching planes take off and land and wondering why his life felt like it was crashing on the runway. Work was going well. He had been at the ministry for four years now and had developed good skills and a good reputation. His love life was up and down, which was nothing new, but somehow the combination of professional routine and emotional chaos left him feeling empty and aimless. "What am I going to do with my life?" he asked himself. What would he become? Suddenly, like the whoosh of a jet taking off for the wild blue yonder, an idea rushed into his head: He would get another scholarship to study, but this time he would fly abroad; he would go to the land of John Deere itself, to the United States. filled with excitement, Milton set up an appointment with a former boss, engineer Fausto Cáceres, who had risen to become an important director in the Ministry of Agriculture.

"I have come here to ask for a letter of recommendation for a scholarship to study abroad," Milton said to Cáceres, sounding much more confident than he felt. Cáceres turned his back to the young supplicant and, without saying a word, drew a piece of paper from his desk drawer. It was a list of twenty-three foreign scholarships for which the government could nominate candidates. "Which one would you like?" asked Cáceres. Milton was dumbstruck. "Don't worry, Milton," said Cáceres, "I know you, and I'll do what I can."

A month later, Cáceres told Milton to pack his bags, he was going to Louisiana State University for a bachelors degree in agribusiness. Milton could hardly believe that his dream was coming true. Elated, he said a prayer of thanks. Even his love life was looking up. He and his girlfriend, Miriam, had resolved their differences, and he felt confident that their love would grow stronger despite the separation. When Miriam saw him off at the airport that fall of 1981, she said softly, "If you are not sure what you feel for me, please write me a letter—or don't write, and I will understand." But Milton knew, almost as soon as his plane touched down in Louisiana, that his dark-haired beauty, so wise and full of faith, was the one for him, forever. And so it was that Milton flew home that December to celebrate the birth of Christ with his new bride. They were wed on Christmas Eve.

In August 1985, Milton and Miriam returned from Louisiana to Honduras for good. With his degree, Milton was guaranteed greater professional advancement. They also brought home a very special American souvenir—their first child, Andrea Alejandra. Upon discovering that his old job in the Ministry of Natural Resources had become highly politicized under a new civilian government, Milton thought that he would move into the private sector. He hoped to become a banker—perhaps a loan officer for agribusiness—and make even more money than before. But before he started his job search in earnest, a friend persuaded him to visit a Christian relief and development agency, World Vision/Honduras, which was looking for a project coordinator. Before he knew it, Milton was offered and accepted the job of disaster relief coordinator for World Vision. It was a period that turned out to be blessedly free of disasters: no droughts, no wars, no earthquakes. Thus Milton and his team turned their efforts toward preventing hunger through agricultural development. In the course of his work for World Vision, Milton got in touch with World Neighbors/Honduras, which was rapidly earning a name for itself in agricultural improvement. He also looked up a man who had bought an uncultivated hillside farm from his uncle several years back and turned it into a training center for peasant farmers. This was the same man who had brought World Neighbors to Honduras: a certain Don Elías Sánchez of Granja Loma Linda.

"Technology is not the answer, Milton," Elías told him, over and over, when Milton visited Loma Linda. "It's only a means to help people develop themselves." The two men would hike through Loma Linda's terraces full of vegetables and past huge compost heaps built in the course of weekend training sessions that Elías now held for groups of campesinos. There were no facilities on the farm, just a hut to picnic in, steep hills transformed into productive agricultural terraces, and a river that fairly sang, the Río Chiquito. Although the granja was brimming with simple, or "appropriate," technologies such as live barriers of grasses to hold soil on the terraces and portable seedbeds made of halved old tires, the real goal, said Elías, was "to get peasants to work with the resources of their 'human farm'—their heads, their hands, and their hearts. That's the real secret of development."

Milton listened intently but found it hard to set aside the revelations of his technology-based education at Zamorano and Louisiana State. He recognized that because of the high cost of inputs, chemical-based agricultural production held little promise for the poor campesino. And he saw much promise in less expensive technologies such as those in use at Loma Linda. Elías, however, was relentless in his criticism of agricultural development driven by technical expertise. "Agronomists are a plague." he would snort, "They strut around with their equipment and their answers, stepping on plants and people as they go." Milton pondered the charges and worked out his response in the course of his job with World Vision. He was gaining valuable management experience, in particular, managing an agricultural project modeled after World Neighbors' and Loma Linda's farmer-based approach to development. He began to spend more and more time visiting farmers, working alongside them both to instruct them and to learn from them. finally, through Elías's prodding and his own discoveries, Milton found his role as an agronomist—getting close to farmers not to develop practices but to develop people.

In 1987, after two "disaster-free" years with World Vision's relief program, Milton faced the prospect of being laid off. In fact, he did so with a certain degree of satisfaction and saw it as an opportunity to start his own demonstration farm, much like Elías's, in another part of the country. But then-director of World Neighbors/Honduras, Roland Bunch, had a different idea. "Why don't you start a clearinghouse of information on cover crops, Milton? We can find the funding, and you can direct the organization." The goal was to help farmers who were successfully preventing erosion and conserving their soil to move a step further and actually improve their soil. This would come about not through the quick fix of fertilizers but through a low-cost, lasting method of planting legumes or "cover crops" such as beans and peas, which fix nitrogen back into the soil, alongside staple food crops such as corn, which leech the soil of nutrients.

In fact, the vision for an international clearinghouse on cover-crop information started with one small bean—the velvet bean—which Hondurans on the north coast had been using for fifteen years to enrich their soil. Roland Bunch was sure that similar practices existed throughout other parts of Latin America and much of the developing world, and Milton was intrigued by the notion of setting up a system to collect, share, and disseminate such information. He viewed the project as largely technical: research, documentation, and information systems. But as usual, Elías stepped in and persuaded him that to have any impact, such an organization must function as effectively in the field, with the farmer, as in the ivory tower and across transmission lines. After six months of thought, preparation, and fund-raising, Roland and Milton secured a $40,000 grant from the Ford Foundation of New York for a 2½ year start-up, and the International Cover Crops Clearinghouse, or CIDICCO, was launched. The agronomist from Zamorano had veered off a well-marked career highway onto an uncharted path that, like the road to Linaca, would be full of both potholes and promise.

Notes

1. Escuela Agricola Panamericana (EAP), *Zamorano, 50 Years* (Tegucigalpa: EAP, 1992).
2. Stephen Schlesinger and Stephen Kinzer, *Bitter Fruit: The Untold Story of the American Coup in Guatemala* (New York: Doubleday, 1982).
3. EAP, *Zamorano*, p. 10.
4. After 1958, financing for the school became more diversified and included non-UFC donors.
5. EAP, *Zamorano*, p. 13.
6. Interview with Milton Flores, Sept. 29, 1992.

4

Preparing the Land

ELIAS SANCHEZ has a face carved with character. The nose, which has done so well in guiding him to farmers redolent with hope, is a truncated beak. His piercing vision comes from eyes set wide and deep under heavy lids and above permanent little pouches, one dripping a tiny mole. His square jaw carries a mouthful of zingers to wake the drowsy and sting the blind. Yet the secret to his considerable talent for communication lies not in his face but under his fingernails. There you will find encrusted the rich, brown earth of Granja Loma Linda, proof of his love for the land he calls "living soil"—a proof that speaks louder than any agronomist's tape measure to the campesino.

When you marry someone, you need to know their inside, not just their outside. When you plant, you must know your soil.

—JORGE AMADOR

At Granja Loma Linda, the secret to farm productivity starts not in the seed but in the millions of tiny organisms that aerate and enrich the soil. And the secret to transforming a farmer's relationship with his land lies in his relationship with the maggot, the worm, the ant, and a host of bacteria and crawlers that decompose matter. Indeed, a farmer who plunges his fingers and arms into the earth and gives it a handshake in partnership with these creatures is unlikely to burn his fields or poison them again.

At Granja Loma Linda, such organisms call home thirty-five acres of steep Honduran hillside that rise on either side of the swift-flowing Rio Chiquito. Don Elías purchased the land, just forty minutes from Honduras's capital, Tegucigalpa, in the 1960s at a low price, as its former owner considered it too difficult to farm. When Elías began to develop the farm in 1980, shortly after he met the Andrade family of Linaca, he was still working full time for the Ministry of Natural Resources. He would simply invite groups of peasant farmers to visit for weekend training sessions, lodging them at night in his house in Tegucigalpa. The fields, on about a twenty-degree pitch, resembled the campesinos' own, except that they were bountiful. Several trainees who went on to transform their own farms with Loma Linda techniques remarked that when they saw Elías's land, they realized that theirs held promise. Elías taught them to make peace with harsh hillsides by preparing the land, and to make peace with change by preparing their hearts to receive new ideas. "Weed your hearts," Elías advised. "If you keep that pollution, that mistake, you won't have room for new joy." Away from the peer pressure of their communities, the trainees learned by doing and were open to innovation.

As in the Association for Coordinating Resources in Development (ACORDE) training program, soil conservation and other appropriate technologies abounded at Loma Linda. Farmers learned to build terraces with contour ditches, which drained rainwater without washing away topsoil. They made compost heaps from farm refuse on each hillside and mixed the organic material with manure to fertilize the small strips of soil on each terrace for planting in Elías's in-row tillage method. They created live barriers of grass and shrubbery to hold the terraces. They watched the patterns of runoff from rains, and where a rivulet formed they would plant a banana tree, azalea, or bougainvillea to stem erosion. They used old tires, cut in half and inverted, as huge planters in which to sprout seedlings for transplanting. They learned to mix natural insect repellents from local foliage. They even planted flowers just to beautify paths through the farm, albeit with skepticism. "Flowers are not useful," the trainees told Don Elías. But he countered that they were "the smiles of the soil." "You have more than one stomach," he maintained, "and flowers are the food of the soul."

Before long, nongovernmental agencies running agricultural training programs in rural Honduras began to stop by the farm, which had no phone, and request that Elías conduct sessions. By 1987, Elías was hiring Fernando Andrade and other protégés to run weeklong sessions for trainees, and in 1989, Elías retired from the government and began working full time at the farm. He built rustic dormitories, showers, and lavatories for men and women trainees, a cookhouse, a dining room, an office, and a sleeping hut for himself. He hired his former housekeeper, Candida Rosa Osorio, who had often helped during the weekend training sessions, to run the cookhouse and administer the training camps. In 1991, he moved up to the farm to live.

Notorious for spurning paperwork, Elías did not keep formal records of his training sessions but relied on Candida and his memory to keep him at the right place at the right time. By 1992, the farm had a phone and a well-defined routine: On Monday mornings, about twenty-five trainees, including campesinos and their extension workers, would arrive at the farm for a six-day training camp, usually in old pickup trucks. The charge for each participant was 75 lempiras, or about $15, a day. The fees, generally paid by a sponsoring third party, covered food for the course and compensation for Elías, Candida, and two to three additional staff who helped with training, cooking, and farming.

The camp marched the trainees through field work in soil conservation, cultivating and marketing produce, and nutrition—80 percent hands-on work and 20 percent lectures, all laced with motivational discourse. Elías and other Loma Linda instructors would draw their explanatory diagrams in the soil with sticks. Lectures took place in the middle of fields or in the rustic dining hall.

Throughout the week, Elías spoke to his trainees as equals, got his hands as dirty as theirs, and emphasized that the tools they were using were their endowment: their hands, heads, and hearts, the tools of the "human farm."

And so it was on a sunny September day in 1992 that seventeen men, six women, and three extension workers from the remote province of Lempira planted themselves in an open classroom under a blue sky, elbow deep in chicken manure. These farmers, part of a rural-impact program sponsored by the Dutch government, wore their best dresses, shirts, and trousers as they worked the terraces of Loma Linda. They considered it an honor to be selected for training and had come a day's journey from one of the most adversely affected zones of Honduras—smack up against El Salvador's border. It was a daring sacrifice to leave their farms for a week, but they came because they yearned to learn and to put new knowledge to the test. A couple of the women had come with their husbands, others were single heads of households, and two were community volunteers. All worked their farms and wanted to improve their land and their lives. A young exchange student from Finland joined them to observe the process.

For Elías—by nurture a product of his mother's backbone, and by experience a proponent of girls' schooling—coeducation served a dual purpose: to train deserving women and to liberate men from machismo. A clue to his thinking on the topic lay on rudely constructed shelves in his little office among agricultural journals and texts. There, a worn copy of *The Second Sex* by Simone de Beauvoir marked a step in his journey toward understanding the situation of women. "It reinforced my thinking," said Elías, "that women don't need to be emancipated . . . . men do! . . . . We have to work with men if we want to help women." By integrating women into his training courses, Elías strove to help men see them differently. By helping Candida and Juana cook and serve, he showed a masculine care that transcended machismo. This process of development changed Vilma Andrade's relationship with Fernando. It helped Candida mature amid oppression. "Men don't think in terms of equality," observed Elías. "Women in Latin America are physically maltreated, and few know who they are . . . . When a girl is born, people are sad. Everyone wants a boy. But women have eyes and ears, hands and feelings, just as men, they have the capacity to develop." Out on the hillside was the proof of it. The women of Lempira, with their clean clothes and hands full of dirt, were providing their kinsmen with a view of their sex beyond the blinders of tradition.

All the while, a former trainee, one Jorge Amador, threw his considerable weight into teaching them the art of preparing the land. First he showed them how to stake a terrace—an old root made a good hammer—and barricade it with handfuls of grasses, which would root themselves to the slope. Then they tested the terrace's ability to hold water using a rustic level—rude slats nailed in an A-frame with a liquid measure poised on the crossbar. The terrace must be constructed so that rainwater flows to the roots of the plant, not off into gullies that would erode the land. Next they hoed a row atop the terrace and carefully removed the rocks, roots, and pebbles with their hands. Finally, they massaged the soil, turning it with organic matter that they'd brought from the compost heap and the chicken yard. What knowledge came with preparing the land? "You have to get rid of the bad elements if you want to grow good fruit," said thirty-two-year-old trainee Juanita Cervantes de Franco, four feet tall and the mother of five.

"What feeds the earth?" asked Jorge, as the farmers broke to return to camp.

"Organic matter," said campesino Lucio Menjivar.

"And what grows with organic matter?" pressed Jorge.

"Living soil!" chorused some women.

"And what does the living soil feed?" The small crowd fell silent, wracking their brains for another agricultural insight.

"Why, us in the end!" exclaimed Juanita, with a sudden grasp of the food chain. "It grows food for us." Indeed, the worms worked hard for Juanita and her five children.

"And how will you share this knowledge?" finished Jorge. "Let's think about that over supper." And so they shouldered their hoes and headed back to camp.

In this way, Jorge walked trainees up to the eight pillars of Granja Loma Linda's methodology of change:

1. Initiate change in small increments—geographically, technologically, and conceptually.

2. Train by doing: Live with and work alongside farmers.

3. Respect human dignity in action and language.

4. Achieve innovation at minimum cost; use local resources in harmony with nature.

5. Achieve all tasks with excellence—no mediocrity.

6. Share what you have learned; ideas unshared have no value.

7. Create satisfaction, both personal and communal.

8. Innovate based on God's wisdom expressed in the Creation, for the process of lasting change is a spiritual one.

Back at the little cookhouse by the river, Candida and her assistant, Juana Cerrato, bustled around a large wood-burning adobe stove. They had already spent six hours that day preparing meals, and now they were frying up potatoes, steaming farm-fresh broccoli, and stirring vats of beans with melted cheese. As Juana stoked the fire, Candida ladled hot food onto platters, poured bowls of fresh cream, and pulled stacks of handmade tortillas from her clay oven, specially built to conserve fuel. The flowers may have fed the trainees' souls, but this surely would feed their stomachs.

Elías, Juana, and Candida served the food in the dining room, where trainees sat around a large rough-hewn table covered in red and white checked plastic, eating, chatting, and preparing for their after-dinner discussion. This was the time when Jorge and Elías hoed the soil of man. "The human farm *produces*," Elías told the trainees, "the physical farm only *reproduces*. You have to grow lettuce and cabbage in your head before you will see it on the ground."

Indeed, Elías's insistence on a holistic approach to farming, involving mind and soul as well as body, began the moment trainees set foot in the camp. He would welcome them, show them their quarters, and virtually order them to wash before lunch. Sometimes he even asked for an embarrassing show of hands as to how many had changed their underwear that day. No places were assigned for meals, and on day 1, extension workers could almost always be found clustered together at one end of the big dining table, prepared to answer for the group in discussions. Elias would make a point of sitting with the farmers and asking them direct questions.

Day 2 would begin at 5:30 A.M., when trainees were awakened to take showers and spend some quiet time discussing the previous day's field training and lecture in small groups. Sometimes trainees would take the initiative and use this time for group Bible reading and prayer. If the campesinos failed to organize themselves, Elías might lead them all in calisthenics. So it was that he affirmed the campesinos' resources but challenged them to apply those resources not only to food production but also to health, hygiene, children's education, and the emotional needs of the family—in other words, to the whole farm. If Elías shocked campesinos into attention with direct attacks on machismo, he also turned class hierarchy upside down by serving them as honored guests. The discombobulation cracked inhibitions and opened minds and hearts to new ideas and to one's own abilities, just as the farmers, in wrenching out stones and turning the soil, prepared their land to receive fresh seed. Although some trainees resented the treatment, almost all acknowledged that it was effective. Jorge was living proof.

That night, before launching into parables of human farming, Jorge bore witness to the process of transformation in his own life of forty-five years. "In 1982, my farm was not doing well," he began, "and someone told me, 'There's a person who can show you how to grow twice as much grain in one hectare. Go to Elías Sánchez, he will train you.'" Jorge recounted how he arose on the appointed day at 5:30, full of misgivings. He would have to walk the two kilometers (1.25 miles) to Loma Linda and be there by 7 A.M. "Should I go?" he asked his wife. She nodded, so Jorge grabbed his brother and they trotted off, dirty from days of farm labor, with no thought of bathing. Elías chuckles when he remembers their first encounter: "We always said that we could smell Jorge long before we could see him." The farming course, Jorge discovered, was called Human Training and Motivation. "When we arrived half an hour late," Jorge said, "everyone else there was clean and well-dressed. My brother left immediately. I had a big breakfast and showered, then attended the afternoon session clean. I have bathed ever since, and I went home at the end of the week a new person."

Jorge explained that after the training, Elías came to his land and lent him some money to develop it. Jorge planted corn, vegetables, and fruit trees using the techniques he had learned at Loma Linda and paid Elías back in two months. Six months later, Vermont-based Friends of the Americas hired Jorge to work on an agricultural development project, and next he began to teach classes with Elías. In 1989, Jorge bought his first car and began to market his own produce: two years later, he installed plumbing and electricity on his farm. Indeed, nine years after his introduction to human farming, with continued encouragement from Elías, Jorge's revelation in bathing at Loma Linda had come full circle. Now he was setting a new standard for hygiene in his community.

Jorge also grew spiritually while he was transforming his farm, giving up drinking and joining the Christian Businessmen's Fellowship. He followed his testimony to the farmers of Lempira with a proverb on drunkenness:

Q: What does man become when he is drunk?
A: A parrot that squawks noisily.
   A monkey that makes the crowd laugh.
   A lion that rises to kill.
   An ass that walks backwards.
   A pig that lies in its own manure.

Q: What is drunkenness?
A: It is a stone. We must get rid of it to cultivate.

During the talk, the farmers regrouped themselves around the big table and nodded in understanding. Shyly at first, they spoke of their problems. Jorge answered them with a parable of hard work:

There was a gringo who arrived at a village in Asia, and the people called him loco because he worked so hard in his fields all day long. But when he showed up at the market, he had huge baskets of produce. People began to respect him. He got a job with the government and became prosperous, and when one day he returned to the village that scorned him, the people had named the park where he slept Parque Loco!

He closed the session with a "story without end" on the family:

Don Pedro lived with Doña Chepen, and they had a boy named Pedrito. Don Pedro was a farmer, and when he was not farming he worked hard making cement blocks. He was very busy and never had time to speak to Pedrito. Doña Chepen woke Pedrito at 2 A.M. each morning to grind corn, and at 5 A.M. he had to milk the cows. At 7:30 A.M. he left for school; at 4 P.M. he came back from school to bring in the cows, and at 6 P.M. he had to help his mother do dishes. Pedrito went to bed at 8 P.M., then got up at 2 A.M. to start again. One day, without speaking to his mother or father, Pedrito ran away. Why did he flee?

Again, the men and women around the table spoke timidly in response. These stories could have happened in their own villages, in their own homes. They had plunged their hoes into the earth that afternoon, and now Jorge was raking their hearts.

5

Planting Seeds

TODAY, MANY organizations are dedicated to preparing the soil of the human farm—to identifying and uprooting environmentally destructive practices from Africa's Sahel to Iowa's cornfields. But most have learned, often the hard way, that to crowd out the weeds of destructive farming permanently, one must plant in their place new seeds of knowledge and a confidence in experimentation.

Such was the quest of Milton Flores as he, with the support of Roland Bunch and World Neighbors, launched the International Cover Crops Clearinghouse (CIDICCO) in 1987, the same year that Elías initiated weeklong training sessions at Loma Linda. Milton first put his energy into acquiring the seeds of knowledge he hoped to sow in the

Before researchers become researchers, they should become philosophers. They should consider what the human goal is. . . . Doctors should first determine at the fundamental level what it is that human beings depend on for life.

—MASANOBU
FUKUOKA
*The One-Straw
Revolution*

If we only plant seeds in the ground, not in people's hearts, then what have we done?

—CAMILLO MEJÍA
Loma Linda trainee

43

minds of campesinos. His research took him from the velvet bean fields of Honduras's north coast to Cornell University's vast agricultural library. He paid a visit to Pennsylvania-based Rodale Farm, a leading international proponent of organic farming and research. While Elías was plunging his hands into the earth of Loma Linda, Milton used his to pull up databases.

Soil improvement was a fertile field of study, and the advantages of cover crops were many: They could fix more than 200 kilograms of nitrogen per hectare (about 440 pounds per 2½ acres) into the soil; add as much as 30 tons of organic matter or "green manure" per hectare to a field; crowd out weeds; produce food for human and animal consumption; and, because they grew during the dry season, protect the soil from erosion. All this they did at a fraction of the cost of fertilizer. Nevertheless, Milton realized that by introducing a technology like cover cropping, CIDICCO risked providing a solution that the campesinos would regard as static, a solution that might become a barrier to its own improvement. "If we are going to be satisfied with promoting the use of cover crops," he thought, "then I don't think we will do more than any other agency."

He mulled this over as he investigated, documented, and photographed examples of cover cropping, procured seeds for experimentation and resale, and set up CIDICCO's information network. CIDICCO started out with a list of sixty-six potential subscribers, and within five years it was reaching almost 500 farmers and researchers in more than sixty countries. When CIDICCO's first newsletter, *Cover Crop News*, went out to its membership in 1991, it featured a benchmark for successful technology transfer: the story of the velvet bean as told by the Honduran coastal farmers who had been planting the legume between their rows of corn for more than fifteen years. "We do not know how the idea got there," the article ran, but "we do know it has been widely and enthusiastically accepted by farmers of the region. Moreover, this technology has not been taught formally by any agency. . . . It has been transmitted spontaneously from one farmer to another." From this and other examples of successfully promulgated innovation, CIDICCO

distilled Loma Linda's pillars of human farming into six tenets for
farm improvement:

1. Start small.

2. Practice what you preach.

3. Respect the local culture; where possible, root innovations in tradition.

4. Work as a servant; share the campesino's conditions.

5. Build local leadership; campesinos must see that they are
   capable of replacing foreign extension agents.

6. Look for spontaneous replication, the most reliable sign
   of success.

The first issue of *Cover Crop News* described with words and
photographs not only how and when the farmers planted velvet
beans, but also at what altitude, temperatures, soil conditions, rain
distribution, and cropping seasons. Readers at home and abroad
could fit the activity to their own context—or reject it as incompat-
ible. Above all, the article listed farmers' own perceptions of advantages, which included not nitrogen fixation but "saving money on
fertilizer," not cost-benefit studies but saving "labor on weeding."
And it solicited readers' experience for future issues, which would
include information on the use of other cover crops: the lablab bean,
jack bean, cowpea, pigeon pea, and choreque. Soon farmers from
around the world would be sharing their experiences. In West
Africa, CIDICCO members at the Benin Research Station reported
that the velvet bean was effectively controlling the spread of weed
grass; in Brazil, the perennial soybean was improving soil and had
become an important source of animal feed; and in Tanzania,
intercropping sunn hemp with corn and bananas was improving
yields. Back in Honduras, farmers working with World Neighbors
had begun using the velvet bean to improve their diet: They cut the
ground, cooked bean with corn dough to make protein-rich tortillas and used the roasted bean as a coffee substitute and, when half-

roasted, as a chocolate substitute. Innovations burgeoned, and they were premised not on expertise but on knowing that experimentation held promise.

To move its work out to campesino communities, CIDICCO produced simple diagrams and a battery-operated slide show that could be shown in rural schools as a means of transferring technology through children. To reach farmers, Milton and his staff—who numbered four by 1991—went out and worked alongside them, spending days in the fields, much as Elías had with ACORDE. Milton knew that he could plant seeds of knowledge by helping the farmers plant a few velvet beans between their rows of maize. The technique was simple: Sow the bean just under the soil with a dibble stick twenty to sixty days after planting the corn. But the process of learning was more complex. "If we just came and said, 'OK, we need to plant these seeds here,'" Milton reflected, "where would the learning be? Farmers know how to plant a seed." Instead of giving answers, CIDICCO staff asked questions: What cropping patterns existed in the community? What varieties of cover crops were known there? What other ones existed? How might this plant or that one react to present cropping conditions? Once farmers thought through these issues, drew conclusions, and made their own choices, they could readjust them as their seeds bore fruit. What was good about a given practice? What should be changed to improve it? "You won't find two villages where farmers are doing the same thing to manage the use of cover crops," observed Milton, "because it is a process . . . it's the way farmers experiment . . . always observing."

The approach—observe nature, experiment, learn, relearn—is akin to that of the man who most inspired Elías Sánchez in his philosophy of agricultural development: Japan's Masanobu Fukuoka, the father of "do nothing farming." In his seminal work *The One-Straw Revolution: An Introduction to Natural Farming,* this highly trained plant pathologist describes a journey he began one day when he noticed a rice plant growing alone in a dry field against all the weeding, fertilizing, and irrigating wisdom of modern paddy cultivation. He began to question all "truths" of agricultural science, left his government research laboratory in Yokohama, and retired to his family's mountain farm to watch nature take its course. There, instead of clearing and cleaning his father's fields, he let the straw of one crop become compost for the next. Instead of planting seeds, he scattered them coated in clay pellets. Instead of killing insects, he let them kill each other, and instead of pruning branches, he let orchards run wild. In his deference to nature, he lost several hundred fruit trees and nearly drove his father to distraction, but in the end he developed a viable natural farm that encouraged the farmer to be like a child—knowing nothing, seeing all.

To Fukuoka, "modern industrial farming desires heaven's wisdom without grasping its meaning."[1] It wants to use nature without understanding it. But natural farming proceeds from the spiritual health of the individual.[2] Today, the Fukuoka farm has become a pilgrimage for farmer-philosophers young and old. Its mountainsides offer cold well water and sleeping huts for visitors, who may stay a while, work the farm, observe, unlearn, relearn. Fukuoka, with his long white beard and black horn-rimmed glasses, lives in such a hut himself and maintains that "anyone who will come and see these fields, and accept their testimony, will feel deep misgivings over the question of whether or not nature can be known within the confines of human understanding."[3] His books, which hold that natural farming leverages nature's wisdom, have spurred thinking and experimentation around the world, even as far away as a farm called Loma Linda, where today 70 percent of inputs are organic and 50 percent of the food on the table comes from the fields.

There by the huts and cold waters of the Rio Chiquito, Elías and Jorge have been planting seeds and building compost piles with the men and women of Lempira. The trainees note where plants hold their reproductive parts—in a fruit, flower, stem, or root. Once they've plucked the seed, the essentials for sowing it successfully are time and space. One must plant after the first rains, with enough warmth and lead time to grow the crop and harvest it before bad weather and cold set in. One must plant seeds deep enough to protect them in germination, shallow enough to allow their sprouts to break through to the light, and far enough apart to give each set of roots room to grow and breathe. Elías, Jorge, and the farmers of Lempira dig their fingers into the earth and plant two kernels of corn at a time, half an inch under the soil and sixteen inches apart to leave room for the thousands of roots of each stalk.

The compost piles, whose organic material is used to enrich the seedbed, are built in niches off terraces that are brimming with produce: large heads of lettuce, broccoli, and cauliflower; fat roots of onions and potatoes; bushes of yucca, manzanita, and avocado, heavy with fruit. On this September day, the sun beats hot, even as leaves and shrubs sparkle with droplets from the night's rain. Plunging his hands in the earth, Elías pulls out a fistful of worms. Shoving them into the compost heap, he brings forth maggots. The students, too, bury their arms in the mound and feel the full heat of nature's furnace—matter decomposing at 80°F. They then build their own compost from basic ingredients: nitrogen, which can be supplied by green matter; potassium and phosphorous, supplied by dry leaves; a yeast, in the form of bacteria-filled manure; and oxygen, which is introduced by interspersing layers of dry material with wet, inserting crumpled trash paper, and poking the mound full of holes. Watering each layer and covering the whole with straw, the trainees leave their masterpiece to metamorphose into rich organic material, which, in the warmth and humidity of Honduras, will take a mere two months.

In the same way, in time and space and with enrichment, Loma Linda trainers plant ideas. They identify seeds of knowledge and sow them deep enough in a campesino's heart and mind to protect her or him from the pecking of skeptics. They sow them close enough to the surface of activity to find their own expression in a farmer's practice, and far enough apart that they can root and grow firm without confusion. Trainers start small—in concept, in technology—and plant their germs in complementary fashion, like corn and cover crops. The ideas are layered, organically, to enrich one another, to conceive and form new thoughts and actions. Under some conditions, the process takes months: under others, years. In this way, Elías cultivates both concern for the soil and care for personal hygiene. In this way, Candida plants ideas for improved nutrition in the minds of women who labor to improve their crop.

Candida's lessons in nutrition take place in a new kitchen annex by the river, a square, squat building with screens on all sides, varnished counters, a modern stove and tables. In a country where the poor have begun buying more Coca Cola and less milk, more packaged snacks and fewer vegetables, Loma Linda finds that nutrition is a hard sell made easier when wrapped in an American kitchen. There, the women of Lempira have been learning to make pancakes, chorizos (a kind of sausage), and a sweet, milky beverage, all from calorie-rich soybeans. The men are invited to join the lessons but decline. According to Candida, the kitchen wall is one of the last to crack in the demise of a campesino's fortress of machismo: "Sometimes you can get them to start helping in the kitchen by making a little cup of coffee," says Candida, "but if they can get away with it, they are happy for the women to do everything." Participant Armida-Lara Escalante, in a bright yellow dress, discusses the value of the new dishes with a visitor, as well as her strategy for introducing them at home:

"Why do you want to introduce them?"

"To improve the nutrition of my family."

"And how will you get your family to eat these dishes?"

"We'll serve them here first for lunch!"

At Loma Linda, campesinos can, without ridicule, try something new: if group approval is won, the practice can be marched home. "I've had trouble writing everything down with my pen," says trainee Camillo Mejía, "but that which is written here," he points to his head, "in my human farm, is there to stay."

Non-campesinos also come to test ideas at Loma Linda. In the evenings, or on Saturdays after a training group departs, friends and neighbors, activists and politicians, professors and practitioners drop by the farm for some of Candida's tortillas and Elías's conversation. Gustavo Alfredo Landaverde, a former member of Congress who has worked with President Rafael Callejas on agrarian reform, is one such visitor. His labors in politics as a deputy from Cortès have convinced him that laws do little to effect change unless accompanied by improved resources and education. He speaks of how for several decades the Honduran government, through the Ministry of Natural Resources, has focused rural development efforts on ensuring that the rural poor are given title to land. The Agrarian Reform Laws of 1962 and 1975 expropriated land that was not under cultivation to redistribute to the landless. But Landaverde and his colleagues see that newly titled campesinos, lacking credit and skills to develop their land, often resell their plots for cash and move to the slums around Tegucigalpa.

A 1992 study produced by the Central American Institute of Development Planning and Economics found that the first thirty years of agrarian reform had benefited only 59,000 families, or 17 percent of the rural population. Meanwhile, 220,000 families, 27 percent of the sector, remained landless or eked out a living on less than one hectare.[4] It predicted a similar result for the ministry's 1990s effort in agrarian reform, called "agricultural modernization." The study concluded: "No agrarian reform can succeed without including an efficient program of education, training, technical assistance . . . financial and marketing services that provide real, sustainable force to integrated development."[5]

Under the agricultural modernization plan of President Callejas, campesinos who sell their land back to the private sector will receive the fair market price. But Landaverde believes that the modernization plan has the same flaw as agrarian reform: The weakest, least educated lose. "The country may profit," said Landaverde, "but individuals, especially the poor, are hurt. Under this law, the companies can buy campesinos' land at a price that they can't refuse. So the reform isn't forced, it's induced, but it basically says that campesinos will never be productive. And campesinos who sell their land don't necessarily invest well. They just lose all food security."[6]

Landaverde, who has been involved with ACORDE, admires how Elías counters such policy with personal efforts to find and encourage the weak. Indeed, Loma Linda has provided thousands of training scholarships to rural poor who lacked financial backing. The Loma Linda approach has planted seeds of recognition of alternatives to reinforce the campesinos' role in the economy. Although the government espouses support for the efforts of grassroots organizations, it remains careful not to tread on commercial interests that feed on chemical farming.

Another regular at Elías's table offers a dramatic example of a weak life made strong through encouragement. Luís Alonzo Morales, called Loncho, literally feels the contours of the farm's beauty when he comes to pay a call. He touches herbs growing in tubs made from old tires, feels the cool wetness of the river run through his fingers, and buries his nose in lilac blossoms near the cookhouse. Loncho sees light with every sense he possesses—taste, smell, touch and sound—but his eyes see only darkness. Elías found him a couple of years back while on an errand in his village. A relative beckoned Elías to come view the tragedy of Loncho's scared, sick household. The boy's father was drunk, his mother handicapped by severe burns; blind Loncho, who took the brunt of his papa's frustration, sat in a corner, huddled and beaten. Elías invited Loncho to the farm and found a school for the blind nearby, where Loncho could learn crafts and independence. Soon Loncho's mother arrived, and the two of them found rooms near Loma Linda where they could start a new life, Loncho making and selling his handiwork. Often they return to wander the farm, share a meal, touch the fruits of nature, and listen to the music of the river.

Not surprisingly, Loma Linda's care for the marginalized has sparked the interest of Honduran human rights advocates. One evening, three human rights workers for Cooperación Técnica Nacional (CTN, or National Technical Cooperation) gather around the varnished table in the farm's kitchen annex to discuss harassment of campesinos by the military. Farmers near the El Salvador border are being taken for guerrillas and harassed, despite constitutional guarantees to Hondurans, foreign residents, and visitors of the rights to life, individual freedom, and safety of person and property. Military personnel in the department of Intibucá have intimidated campesinos into selling their land and their trees for a fraction of their value. Although organizations like CTN, which helps campesinos conserve land and protect forests, work to inform them of their rights, the farmers' resolve frequently crumbles in the face of men in uniform. In Elías's view, only empowering campesinos to take responsibility for their own education and development can bring them confidence in their rights and thus shield them from abuse.

One campesino community that has made the leap from taking it on the chin to taking responsibility is El Espíritu, in the department of Copán on the Guatemala border. Francisco Salinas, agricultural project coordinator for Catholic Relief Services (CRS) in Honduras, which sponsored El Espíritu farmers' training at Loma Linda, tells their story. In 1987, when CRS agreed to back an integrated agricultural development project in the zone, the community's baseline statistics were dismal: 27 percent infant mortality, 72 percent child malnutrition, 60 percent adult illiteracy, 70 percent deforestation and 30 percent of villages inaccessible by road. But the residents of El Espíritu possessed one resource that many communities lacked—faith in one another and in their Creator. As participants in the Catholic lay movement, Delegates of the Word, they had for twenty years been working together, sharing concerns and care for one another. They had created a context of trust and experience of personal development that gave them confidence that they could, with some knowledge and effort, improve the quality of their lives. With CRS came the means: with ideas gleaned from Loma Linda, World Neighbors, and others came the inspiration.

The project began by planting and watering just a few seeds: training fourteen men and five women from El Espíritu in soil conservation and improvement, nutrition, health, and hygiene. The trainees brought these lessons home to their neighbors on foot and on horseback in regular personal visits, in group workshops, church meetings, and songs. The regional statistics may still hide their progress, but in village after village, change is afoot. In one, Catalina Ramirez convinced her father to stop burning fields and subsequently increased family corn yields fivefold. In another, José María Márquez Castro transformed his once-abandoned hillsides into cascades of productive corn terraces. Luís Acosta Alvarado, working to promote health and nutrition, observed infant deaths in his and surrounding villages fall from between six and ten a year to just one or two.

The project participants have designed a logo for themselves that shows the basics of their new accord with nature: Within a gold circle, a yellow sun rises over the crest of a green, terraced hill. On its terraces, a farmer prepares strips of soil with his hoe, his pick, and a rude A-frame level: in the distance, a rain cloud hovers. It represents, as the project calls itself, "A New Dawn in the Countryside." COSUDE, the Swiss government's aid agency, has captured the lives of the men and women from El Espíritu in a handbook of photographs and testimony. There you will find a list of ten mandates for the extensionist, which articulates the pillars of human farming.[7] They are the same tenets for change taught at Loma Linda. But if you ask the extensionists of El Espíritu, they will say that they are their own.

Notes

1. Masanobu Fukuoka, *The One-Straw Revolution: An Introduction to Natural Farming* (Emmaus, Pa.: Rodale Press, 1978), p. 118.
2. Larry Korn, "Introduction," in Fukuoka, *One-Straw Revolution,* p. xxv.
3. Fukuoka, *One-Straw Revolution,* p. 29.
4. Efrain E. Diaz Arrivillaga, "Un Breve Analysis de Ajuste en el Sector Reformado," in *Honduras: El Ajuste Estructural y la Reforma Agraria,* eds. Hugo Pino and Andrew Thorpe (Tegucigalpa: POSCAE,1992), p. 114.
5. Miguel Morillo, "Los Campesinos, la Reforma Agraria y el Ajuste Estructural," in *Honduras: El Ajuste,* eds. Pino and Thorpe, p. 128.
6. Interview with Alfredo Landaverde, Sept. 1992.
7. Pascal Chaput, *Un Neuvo Almanecer en El Campo* (Tegucigalpa: COSUDE, 1991), pp. 8–9.

6

Tender Shoots

FERNANDO ANDRADE, Jorge Amador, Loncho Morales, Candida Osorio, and Milton Flores all share an unusual bank account. Its deposits are in time, care, and training; its original benefactor is a short, feisty Honduran, Don Elías. Strangely, the account generates interest for donors and recipients alike, enabling all to set up new accounts for others. Says Don Elías, "We don't teach here, we share information in two directions." In this way, the riches of Loma Linda are credited to communities near and far. Sometimes Elías himself takes the training out from Loma Linda to the field; more often, a sponsoring non-governmental organization (NGO) provides encouragement to Loma Linda trainees back in their communities. And the trainees, in turn, train

Model farms have a role motivating farmers to try something new and helping them to realize there is a whole world out there of better agriculture. . . . Whether farmers can recreate that world themselves is another issue.

—ROLAND BUNCH
COSECHA

You cannot change behavior with a one-week course . . . If you don't provide follow-up, as soon as these people go back, one chicken can destroy the whole thing.

—DON JOSE ELIAS
SANCHEZ

others. Without this follow-up, growth arising from a campesino's initial investment in Loma Linda training could soon be squashed or squandered.

Fernando Andrade knows well the value of such support. When he first began to experiment with the techniques he learned through the Association for Coordinating Resources in Development (ACORDE) training, his neighbors levied harsh criticism. "If you grow up in a community where people throw trash in the street," observed Elías, "you will be criticized for picking it up. The more Fernando worked to leave the club of soil destroyers and become a soil conservationist, the more he was attacked." What did he do? "He closed his mouth and did a good job," said Elías, who gave Fernando much personal encouragement at the time. Later, Fernando took another step to ensure that he would have space to grow: He transplanted himself and his family.

In the fall of 1992, the trainees from Lempira have gone home, and Fernando and Vilma have come from Linaca to Loma Linda to help Elías with another group of farmers. Fernando and a young farmhand, Jaime Delgado, are transplanting onions from seedling terraces that rise from the Rio Chiquito to new terraces carved out by recent trainees across the river, past the compost heaps and behind the dormitories. Transplanting is a delicate process that demands care not only in uprooting a seedling from its seedbed but also in reestablishing it in new soil. The onion seedlings at Loma Linda are growing cheek to jowl, about 4,000 to a terrace in their seedbeds, each plant bearing several leaves. After transplanting, they will be spaced fewer than 100 per terrace.

Jaime and Fernando lift the plants carefully from their beds, carry them across the river in bundles, and replant each root two hand-widths apart. They have prepared strips of soil on the new terraces just as they have taught trainees: out with the stones, in with the organic material. Jaime has used a thin hoe to till strips on each terrace for planting, turning the soil with a mixture of chicken manure and fresh compost. Fernando pushes two fingers down deep in the soil to hollow out a space for each onion root. He makes sure that the ground around the space is level, so water will not wash in, collect, and drown the root. In goes the onion; then weathered fingers gently replace the soil around the bulb. Today, the little onion plants look tiny, lost, two hand-widths apart on a wide terrace. But in a few short weeks, they will grow into their new surroundings.

So it goes with each of us when we outgrow our environment. So it goes with Honduran hillside farmers. Our growth potential, like the onion's, rests in the *way* in which we are transplanted, not in the act of transplanting alone. Do we wait until our roots are mature enough to lift from the soil without tearing them? To travel without wilting? Do we time our move in keeping with our growth? Prepare a new space to reroot? Allow hands to lift us, help us, plant us upright, so rains won't drown? To pack in support, even new responsibilities? "When you transplant," says Don Elías, "You don't transplant with your head—you transplant with your heart and your brain. You make a little hole, and you place the plant with love."

In the case of Fernando Andrade's family, they first felt the need to transplant themselves, back in 1978, out of weakness. Their land was barren, their children were hungry. But the strong hands of Don Elías pushed them to reconsider, to help heal their land and to care for their children in a new way, strengthening their roots. Later, after reinvigorating their farm, the Andrades were strong enough to transplant themselves in a way that allowed them to share their lessons with hundreds of other campesinos and grow in their own knowledge of people, places, and processes. But the preparation, the process of strengthening their roots to support such a move, came with follow-up, back in Linaca, from Don Elías and others.

After the excitement of the ACORDE training sessions and the visits to Guatemalan hillside farmers, Fernando continued to make compost and work hard to improve his soil, because he saw results. Elías left him alone for a while, then stopped by one day and saw that he had made tremendous progress. Outside Fernando's house, his crop had increased eightfold to 80 quintals (8,000 pounds) per manzana (1.7 acres). Inside his house, family relations were changing for the better. Whereas Fernando's prior farming knowledge did not exceed the lessons his father had passed down, he now was growing in his own wisdom, gaining confidence in experimentation. Observed Elías, "He began to believe in himself. He started to show signs of living. He even changed the concept he had of fathering and became more of a friend to his daughters."

Indeed, Fernando's own intellectual growth made him consider his children's education. "I began to wonder whether I should give my children an opportunity to study. I became convinced that they should not be the same as me," said Fernando. Following his new conviction, he sent his daughters off to school. Vilma too was drawn into the new cycle of learning. When Elías first began the ACORDE training in Linaca under the Ministry of Natural Resources, he was careful to open it up to farmers' wives. "If your job is to cook," he told them, "then you need to make sure you can grow good food." Elías had worked with Vilma and other Linaca women, beginning with techniques as simple as hand watering the fruit trees in the yard, then teaching them to plant vegetable gardens. "I began to work a bit of the yard," says Vilma. "I watered the orange trees, I planted lettuce, onions, celery, beets, sweet chilies. I worked with my daughters to prepare the soil and plant the seeds."

One day, Elías dropped by while Vilma was grinding corn, bent over, pounding the kernels with stones. He showed her a new way to grind so that she could keep her back straight. "I worked with much more enthusiasm," says Vilma. "Another teacher [from the Ministry of Natural Resources] said that he would come by some afternoon and teach me more. He showed up and taught me to classify my vegetables for sale, onions, cabbage, and lettuce. Then he asked if I would like to do macramé." Vilma finished the macramé class, and another promoter volunteered to run a nutrition class, where Vilma and her neighbors learned, like the trainees at Loma Linda, to cook with soy and to make buttermilk and soy milk. With the help of their teachers, the Linaca women organized themselves to take baked goods, garden produce, and macramé plant hangers into the market and sell them. For Vilma, the process was empowering. She had more knowledge, more responsibility, and, as a result, more say in many of the family decisions. "This land was my personal land," says Vilma, of the Linaca homestead. "But Fernando knew more than I did, because he had worked on a farm. I had to learn."

The Andrades' roots in Linaca grew and strengthened. Fernando became the president of a Linaca improvement society. He, his brother, and their neighbors petitioned the government for running water, electricity, and an improved school, all of which they eventually received. At home, Fernando enlarged his kitchen and built a fuel-saving Lorena stove. "My first thought, when I began these changes," admitted Fernando, "was to increase my own production—and I did. But then, I felt a growing desire to help other people in the same situation as me to solve their problems." That desire, and his success with his own farm, led Elías to recommend Fernando as an agricultural consultant to a Honduran development project sponsored by a church in the Pacific Northwest, the Lincoln Christian Community. The project, called Tierra Nueva, aimed to train 1,500 campesinos over three to five years in Minas de Oro—the region where a young Milton Flores had first fallen in love with farming.

Fernando got the job, and the family transplanted itself. Slowly and surely, the man who had trouble looking people in the eye found himself speaking in front of groups of farmers, planning lessons, instructing others. "During these years I was working with my head, my hands, and my heart," said Fernando. "This program wasn't just training people in agriculture, it was training them in human development and family integration." The Lincoln church had sent down a young American couple to head the project, but Fernando was the agricultural training expert. "At first the work was problematic," he admitted. "Elías wanted the program to be connected to the Ministry of Natural Resources, and our position was that it should be the people's program. In the end we decided that the program was neither the ministry's nor the Lincoln community's, it was the people's."

As a trainer, Fernando found himself transformed yet again. He saw clearly, as Elías had underscored, that development training *was* a process of good ideas displacing bad ones, with information shared in two directions. In the course of his service with Tierra Nueva, he developed a strong spiritual motivation for his work and, along with other members of his family, entered into a new faith in Christ. As his faith deepened, he became more aware of the behavior he was modeling to others. "You know, in Latin America there is a lot of machismo," said Fernando, "but I sold my pistol—it can't protect you. Only God can protect you. Now I carry a Bible."

The eight steps of human farming—the cycle of transformation from trainee to trainer, from initiating change to sharing what one has learned, with innovation based on Creation's wisdom—were taking place by the late 1980s not just in Minas de Oro but throughout the highland departments of Honduras. In many cases, the cycle began at Loma Linda, with follow-up in the field carried out by NGOs or government extension. By that time, Loma Linda had earned enough of a reputation to acquire its share of critics, those who found a one-week training session too superficial to effect lasting change. Over at CIDICCO, Milton Flores listened intently and defended his mentor. "You don't take people to Elías Sánchez to perform a miracle," he said. "The idea is to shake people up and plant seeds so that they realize they can do things they never dreamed of. We all need follow-up, but we also need something to motivate us. Sometimes I compare the Loma Linda course to Dale Carnegie . . . it's like an injection of enthusiasm . . . a spiritual retreat. But nothing replaces follow-up and what you can do in the long term. That's why it is so important that Loma Linda works with other agencies."

Elías himself says that Loma Linda's effectiveness lies in the fact that it does not take responsibility for all the steps in the human farming cycle. "In the Bible, John the Baptist's followers pointed to Christ and said, 'Hey, he's taking all your people!' And John replies, ' . . . Joy is mine. . . . He must become greater. I must become less' [John 3:29b–30 New International Version]. That is exactly our motive. If I want [a trainee] to grow, I have to decrease, to accept that he or she has the right to increase, to learn to take responsibility. I give you significance if I *listen* to you."

One of the key agencies with which Loma Linda linked forces at the outset was World Neighbors (WN). In some cases, WN brought campesinos to Loma Linda for a week of inspiration, then followed up when they were back on their farms. In others it conducted solely field-based training, as had ACORDE. Roberto Zepeda, a campesino from the department of Comayagua, was one of many to benefit from WN's extension work. As in the case of Fernando, Roberto's father had pulled him out of school after first grade and put him to work on the farm. Roberto didn't mind. He truly liked agriculture. He was one of seven sons, and his father needed all of them to work the land. As a teenager, he decided that he would like to marry, and at sixteen he began preparations by farming his own plot of land apart from his father's. At nineteen, he married Aida Luz, a young woman from a neighboring village. "It was a small sacrifice to go outside my village," he says with a wink and grin. They worked hard on the hillsides, growing corn, beans, cane, and platanos. The soil was not very fertile, but it produced enough.

Ten years later, when Roberto and his wife were twenty-nine, and had two children in tow, a WN extensionist stopped by their house. Roberto had never heard of the organization, but the extensionist explained its objectives, and Roberto joined the program, which entailed meeting with other farmers every two weeks to discuss soil conservation. After reflecting on the discussions, Robert made another sacrifice—this time a big one—and began changing his customs, adopting terracing, in-row tillage, and other soil conservation techniques. His traditional methods were producing just enough to feed the family; if the new methods did not work, they would go hungry. "It was a bit dangerous to change, but I kept with it," he said. His yields improved. "Then we had a course on soil conservation and health, and my wife and I were both invited."

After a year of on-site training in 1986 with WN, Roberto kept working his land and began trying to share his ideas with others. National and international interests listened. "A lot of people asked what I was doing," he says. "Groups came to visit from the Ministry of Natural Resources in Honduras and from CATIE [the Tropical Agricultural Research and Training Center] in Nicaragua.[1] We were a real tourist attraction." They invited visitors into their home, and Aida Luz showed them her vegetable garden and explained how she grew cabbage and spinach to make sure that her children received balanced meals. The Zepedas began measuring protein, carbohydrates, and vitamins in their daily fare and switched from drinking sodas to fruit juices—papaya, lemon, lime. Above all, Roberto became comfortable talking to others. "I can communicate now," he says. "I used to be very shy. More than anything, now that I have older children, I can understand them." The informal education convinced him to provide his own children with a formal one. "My oldest is very motivated and may get to college," he said. "With the others it's the same."

A year later, WN recommended Roberto for a job as one of its extensionists. He ensured that other family members could look after his farm during the workweek and accepted the post in El Jute, two hours by bus from Comayagua. Roberto became a trainer of farmers a year after being a trainee himself. The campesinos were somewhat resistant to new ideas, but he understood their distrust and let his yields do the talking. Sometimes the distrust was between neighbors, even family members. Indeed, the men of El Jute carried guns, and the cemetery was full of people who did not get along. At times, two members of the same family could not attend a community meeting because the animosity ran so deep.

And yet, the forefathers of the farmers of El Jute had passed down sound building blocks for soil protection and improvement. The traditional *milpa* system intercropped corn and squash, providing the food security of a dual crop. The squash, a cover crop, was not demanding of nutrients, and its leaves turned to mulch. The practice had nearly disappeared, and the road to rebuilding trust in positive traditions and between kinsfolk was a long one. After four years—with terraces and contour ditches beginning to line El Jute's ridges, the practice of intercropping spreading, and community relations improving—the local farmers said good-bye to World Neighbors' formal presence in their village, but they did not say good-bye to Roberto. In 1991, he was hired to open CIDICCO's first field office to train El Jute soil conservationists to become soil improvers.

Milton Flores also had found that on-site followup was a key to lasting change, and he was moving CIDICCO's research from its newsletters out to local farmers. As had been the case in the Andrade, Amador, and Zepeda families, El Jute farmers' new attitudes toward the land and their livelihood led some of them to new attitudes toward health, education, and equity in the family. But not all villagers were convinced. "One El Jute woman participating with CIDICCO was very concerned that people were forgetting education," recounts Milton. 'Agriculture is more important to *you*,' she said. 'but education is more important *really*.' She thought that education was teaching kids to read and write, but it's much more than that. Agricultural development is teaching and learning human development."

Nevertheless, Milton realized that in the business of educating farmers who had no formal schooling, classroom materials could fill in some gaps. A globe, for example, was a powerful tool. "The questions: Where are you? Where are you standing? Where are you living? El Jute, Honduras. Where is Honduras? What are relations between crops and the moon and sun and eclipses? This type of approach is very good for teaching adults," says Milton.

Ironically, during this same period, Elías had found a group of formally trained agriculturists who wanted some nonformal training—the *agronomos* of Zamorano. Elías's former supervisor at the Ministry of Natural Resources, Rafael Diaz, had returned to his alma mater, the Pan American Agricultural School, as a professor and was teaching a course in hillside farming. The agronomy students of Zamorano began visiting Loma Linda each Friday to observe the fruits of organic farming. "We started the visits back in 1983," said Diaz, who later directed World Neighbors in Honduras and today works for CARE. "It's a really good thing that Zamoranos get to see what Elías has done, because it is so different from what they see at Zamorano. He's telling people you can make the changes in your life by yourself.

You only have to believe in yourself. He's showing them people development instead of farm development."

Raúl Zelaya, an instructor and administrator at Zamorano, tells how initial efforts to expose students to alternative farming grew to a research specialization. "The school's principal mission is to form technical agronomists for Latin America," says Zelaya, "but there have been people interested in social development." After the hillside farming classes began, faculty got together and wrote a proposal to the Kellogg Foundation. "We were awarded $1.3 million [in 1988] to start a rural development program. In the fourth year [1992] we ... obtained another grant of $1.5 million." Zamorano's W. K. Kellogg Center and Rural Development Program moved extension services out to communities, working with farmers in four zones (including Tatumbla, right next to Linaca) on experimental vegetable production, latrine construction, small animal husbandry, grain storage, and pest control. Moreover, rural development became an optional specialty for engineers in the four-year agronomy program.[2]

"When I said 'extension work' for the first time it was in 1988," says Zelaya, who directed the Rural Development Program until 1992. "I asked which students would be interested in learning this, and one out of 166 raised a hand. Now," he continues, "there is a group that equals 25 percent of our four-year class that call themselves the 'developers.' About a third of the three-year class is interested. The whole idea of this development course is not to prepare extension agents. We want to prepare technical agronomists. We want to prepare entrepreneurs, but with a better understanding of what constitutes development and extension of adult education."

On Fridays, Zamorano's future entrepreneurs arrive at Loma Linda and wind their way up the hillside, over footbridges made of old tires, and out onto the terraces, where trainees are working with Jorge. This Friday, there are twelve Zamoranos, including two young women. They come from El Salvador, Guatemala, Honduras, Bolivia, Costa Rica, and Ecuador. There is even an exchange student from North America. Elías explains to them the benefits of organic farming and has the students shove their hands into compost, acquaint themselves with the worms and maggots. They observe the intercropping of fruit trees and vegetables, cover crops and corn, and they spray mixtures of natural insect repellent on their leaves. Then Elías marches the young elites over to meet the campesinos.

"Do you have any questions for each other?" asks Elías. Silence. "OK," he says to the *agronomos*, "I'll give you their two pieces of advice: Don't smoke and don't step on the plants!" Everyone laughs. "Right?" Elías asks the campesinos. "You see, they are just polite. Campesinos take more care in their fields than agronomists, because they don't have lots of money to waste." The agronomos, in their blue jeans, blue work shirts, and boots, with tape measures strapped to their belts, grin sheepishly. "Most of us want to go home and run our own farms," admits nineteen-year-old Mauro Mendizabal, a second-year agronomy student from La Paz. "I know it sounds selfish—we should help others. But most of the students do have quite a bit of money. That's just the way it is." Mauro and his classmates turn and follow Elías down the hillside single file, with the precision and discipline of Zamoranos. Suddenly, Mauro breaks out of line and shifts left. In his path was a tender green shoot.

Notes

1. The Tropical Agricultural Research and Training Center, known by its Spanish acronym CATIE, is based in Turrialba, Costa Rica, with satellite operations in Nicaragua.
2. The Pan American Agricultural School offers a three-year agronomy degree which graduates about 150 students a year, and a four-year agronomic engineering degree, equivalent to a bachelor of science, which graduates about sixty students a year. It also offers a five-year agronomic engineering degree with thesis.

Granja Loma Linda

*Don José Elías Sánchez,
the teacher, in his open
air classroom*

*Granja Loma Linda
on the Rio Chiquito*

*Don Elías preparing
the land with trainees
from Lempira*

Loma Linda Technologies

*Culverts made
from old tires*

*Trainer Jorge Amador
planting seeds*

*Elías shows visiting
agronomists some
tire technology*

*Planters made
from old tires*

*Jorge shows off
compost heap
made by trainees*

*Terracing with in-row
tillage and agro-forestry*

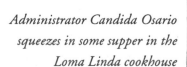

*Administrator Candida Osario
squeezes in some supper in the
Loma Linda cookhouse*

Human Farmers

Three generations improve the Santos family farm near Soccorro.

The Andrade Family

Sonia, Fernando, and Juanita Andrade with their 10' corn in Linaca

Fernando, Vilma, and granddaughter Jili Michele in their backyard

The Flores Family

Miriam, Milton, Mauricio, and Aaron in Tegucigalpa

Andrea at 8, as her family remembers her

Roberto Zepeda and Milton Flores, of CIDICCO, take stock of an insect infestation

Two Different Study Halls

The Pan American Agricultural School in El Zamorano

Zamoranos *visit Loma Linda on Fridays to learn about organic farming*

7

Storms and Vermin

INSIDE Λ two-room house that Roberto Zepeda built in El Jute hangs a large poster that warns of the effects of slash-and-burn farming. Photos depict skinny cows, arid land, and malnourished children in fields where trees have been destroyed. Another poster shows the effects of soil conservation and cover-cropping methods: Fields with terraces buttressed by stone walls and cultivated through in-row tillage sprout tall, lush, green stalks of corn. Outside Roberto's house, the contrasts are rarely so stark. Today, his experimental plot—terraced, tilled, and protected with organic insect repellent—is alive with vermin. The live barriers of grasses planted to hold soil are vibrating with butterflies, grasshoppers, spiders, beetles, and other insects. Hundreds of black crows sit sentinel on stalks of corn bent to dry, oblivious to a scarecrow flapping its shirtsleeves in midfield. Worms have chewed holes

> *I have to refer to your proposed topic for Chapter 7 . . . for there is no more appropriate time for me to talk about storms. . . . My wife . . . and kids find ourselves going through the most devastating experience of our lives.*
>
> —MILTON FLORES
> fax to author,
> September 7, 1992

through fat heads of Chinese cabbage, and a plague is threatening the peanut crop. Milton Flores, who has come up from Tegucigalpa to join Roberto for a day in the field, pulls a yellow beetle off a cabbage leaf. "We have the whole gang here!" he smiles.

In a campesino's calculation of risk and return, breakeven equals subsistence survival. His or her only incentive to risk innovation is visible success—sowing radishes that sprout in two days, cilantro that matures in three weeks, all commanding good market prices. But failure lurks near every innovation in many disguises: bugs, disease, bad weather, market imperfections, and, above all, family strife and community discord. Roberto spreads lime on his tomato crop to ward off the bugs and points out that some of the insects are good—the spiders are his coworkers, eating other predators. He is experimenting with intercropping a foul-smelling weed called *chinche* (after its foul-smelling namesake, the bedbug) to repel other insects and is using different organic compounds to stymie disease. He has observed that insect infestations come and go at different times of the month when the moon is at different stages, and he is doing his best to anticipate and forestall invasions. Do it yourself and see if it is worth the effort, he says. Create evidence that what you advocate can be done. In contrast to local architecture, Roberto has cut sizable windows in the walls of his home to allow air to circulate. And he has built a two-sided roof, which is less costly than traditional four-sided roofs. He is spending his first year with the International Cover Crops Clearinghouse (CIDICCO) quietly planting, building, fighting storms and vermin, and villagers are watching. Some have begun to plant carrots for the first time; some have asked him about his potatoes.

Elías Sánchez often notes the trouble Honduran extension agents have putting themselves in the shoes of campesinos. Imagine, if you will, that the game has been moved to North America. You are at home on a Saturday, in Ourtown, USA or Canada, and a well-meaning stranger comes to your door and says, "Hey, I know you watch TV all the time, and it is very bad for your morale, your muscle tone, and your community spirit. I want to help you start sports teams and community activities. I want to help you work together with your neighbors!" You might tell the strange enthusiast that you are happy watching TV, thanks all the same; the bills are getting paid, and he's interrupting your favorite program.

But suppose the stranger said that he could help you double your income and your spare time with some tips for doing your day job more efficiently? Perhaps your ears would perk up. Perhaps you would listen when he suggests that you use some of that increased income to improve your home, your garden, and your children's education, to buy bicycles for family exercise. If the secret to working more efficiently is, in fact, to work with your neighbors, then you might seek their collaboration. This example may seem ridiculous, but the point remains: Even among populations where the risks of innovation are fairly low and the rewards are apparent, it often takes fairly high market incentives or assured improvements to quality of life to change patterns of behavior. In El Jute, people are poor, tired, and downtrodden; they are just making it. What are their incentives to risk change? Why should they trust a stranger in their midst?

There are thirty-five families in El Jute, and sixteen are participating in the CIDICCO training program. "We've spoken to people, and sometimes they don't want to participate," says Roberto. "People tend to be distrustful of the unknown. They have other things to do." But even the nonparticipants take note. "The advantage is that we have better corn and beans, and we've started growing vegetables smarter. We plant the corn closer, and it's less labor to water and to weed," says Roberto, "but it takes time for [our neighbors] to gain motivation." The fields next to CIDICCO's are terraced, and their owner has hollowed out a smooth, semicircular contour ditch to drain off rainwater. The acrid smell of pesticide hangs in the air. Roberto crinkles up his nose in displeasure. Clearly the farmer has bought into soil conservation but is losing faith in organic farming. "The problem is that people always want to see beautiful fruit," says Roberto. "They don't care about the inside. Organic fruit may have a hole here or there, but it is better." You have to taste it.

Farmers have to taste the fruit. Development workers have to taste the hardship of the farmer. Roberto, even as he fights the insect infestation in El Jute in October 1992, learns that his family farm two hours away in Comayagua has been devastated by floods and landslides. He'll have to go home early this weekend and dig it out. The unseasonably heavy rains have caused landslides as far south as El Jute. Moving across the fields, Roberto climbs a hill, up terraces of lettuce, through a field of corn, to two compost heaps belonging to farmers in the program who have joined their fields and *are* working together. Their land abuts that of a farmer who has chosen to remain outside the program, and it offers a view that sharply juxtaposes soil conservation and soil erosion. Across from the CIDICCO farmers' neat terraces of celery, cilantro, and corn, a steep hill planted with maize falls away to a small valley. Its scraggly stalks are widely spaced, and its soil is washed away, exposing large rocks that jut out between the corn.

Even the farmers who have joined the program, who have increased and intensified production, who have evaded pestilence, face tremendous hurdles to realizing their gains. Milton Flores finds the campesinos increasingly interested in talking not about production but about strategies. "They are realizing that they need to plant different crops. In the past, if one planted cabbage, everyone planted cabbage." When all the cabbages hit the market, the price would fall. Now farmers want to diversify to obtain good prices. Much as they have come to see their families' health, education, and welfare as part and parcel of a successful farm, they are beginning to look at agriculture as a whole process, from production and pest control to marketing.

There is a bus that goes to Tegucigalpa each Saturday, and it takes farmers and their produce to a special market created by the government to allow campesinos to sell directly to consumers. If you arrive at the farmers' market early, you will see the bus unload its cargo of campesinos and watch middlemen, or *coyotes*, swagger over, surround the farmers, and buy their produce in bulk with little discussion. The *coyotes* then turn around and double the price to customers in the market, customers who will haggle for five minutes over a difference of 10 centavos. "The farmers know that this is happening," says Milton, "but they are ashamed and afraid to place themselves in the market and sell." A farmer named Pablito, who gave CIDICCO the land in El Jute for its experimental farm, never ventures to the market. He sends his wife. "They are afraid," says Milton, "that if they offer [vegetables] at a certain price and someone says, 'No, that's too expensive,' they will get embarrassed, discouraged. It's psychological."

José Benito Ponce, fifty-two, and his son Andrés, twenty-four, are exceptions who may change the rule in El Jute. Six years earlier, they had a chance, under the World Neighbors extension program, to visit Granja Loma Linda and see what could be done with creative thinking, hard labor, and care for the land. They also listened carefully to Elías's lectures on marketing produce. "It gave us lots of ideas," says Benito. "Before we used to work ten plots of land; now we concentrate our efforts." Indeed, the Ponce family has *reduced* the amount of land it cultivates from four manzanas spread over ten different plots to one manzana (1.7 acres) in one large plot. They have dedicated a portion of this intensified agriculture to commercial varieties. Unlike the Andrades' Linaca farm, the Ponces' land was fertile; but like the Andrades, the Ponces had reached a dead end in improving it. "We had good soil," says Benito, "but we didn't have ideas." Now Benito and Andrés have a compact farm with terraces of closely planted corn and cover crops, vegetables, pasture for their workhorse, and a drained lowland field planted with cilantro, a cash crop. "We never cultivated this part of the farm before," says Benito. "It was abandoned." Now, with contour ditches, it has become his most profitable field. Benito and his son braved the marketplace themselves in 1992 to sell their cilantro crop and received a good price of 80 centavos, or about 16 cents, a bunch.

Milton and Roberto crouch down and work with Benito and Andrés, harvesting cilantro and tying it in bunches. Milton always works with the farmers when he visits. He picks up a hoe and tills for an hour or two. "It's a relaxing way to discuss successes, problems, new ideas," he says. Milton feels especially close of late to the needs and fears of the men and women of El Jute. They have helped him through a storm whose clouds still hover around him and his family, black and heavy. As Milton leaves El Jute, a farmer beckons him into his home. It is the farmer who owns the beautiful, productive terraces overlooking the eroded hills. His house is freshly painted a bright turquoise; his yard is clean, and his chickens are kept safe from scavengers in a large wire pen. It is not hard to see the connection between caring for one's land and caring for one's home.

Milton emerges from the house with a sweet-smelling gift: a bag of piping hot corn tamales, made from the first harvest of the farmer's maize crop. Driving out from the neat home, under plum trees packed with young boys picking fruit, Milton muses as to the real impact of World Neighbors' and CIDICCO's work. The quantitative results will be straightforward. He and Roberto will document and correlate the use of cover crops with changes in farm yields. But the qualitative results will be important to ensure lasting change, and these are much trickier. Will the group of farmers in the program ever become an engine for change in the village? Will they reach step 6 of Loma Linda's human farming methodology and share what they have learned? Just imagine, if two people working together with just half an acre each would share their vision and skills with two others, then that same year, four people would be practicing soil conservation on 2 acres. And if those four each taught four more the next year, and those sixteen each taught sixteen the next, and replication continued apace, by the fourth year, 65,536 farmers would be conserving 32,768 acres. "That's a big impact on rural land conservation," says Milton. CIDICCO is trying hard to foster such exponential change.

Farmers in the program already have a platform, based on the respect they have gained from the community for their economic successes, to promote social improvements in El Jute: schools, education, potable water systems to replace two open wells. They could put their creative thinking to the service of the community. They could share what they have learned. They could work together. And yet, for the most part, they do not. Just recently, a few farmers asked Roberto to arrange a truck for them to carry a load of manure for their composts. The next day, a few more farmers asked for the same service, and then a few more. "Why don't you get together and rent one large truck for everyone?" Roberto asked them. Even after several years in the extension program, they lacked the reflex to share their burdens.

Driving back to the capital, Milton passes a green meadow dotted with white crosses and flower-filled plots on the outskirts of Tegucigalpa. His face darkens with a memory that brings pain. Some minutes later, he stops at a home as compact as the Ponces' farm. It is the Flores home, a cooperative town house on one of the half dozen hills of Tegucigalpa. In front of the house is a little garden with old tires filled with earth and sprouting corn stalks. There are two coffee bushes and a little eucalyptus tree that Milton's eight-year-old daughter, Andrea Alejandra, helped plant. Inside the house, three bedrooms, a bathroom, and a small kitchen surround a family room for sitting, dining, playing. Aaron, six, and Mauricio, four, are using the space to full advantage, tumbling over each other like kittens, laughing and stacking their Lego blocks. Behind them on a coffee table sits a picture of their sister, born in America. Her head is cocked to one side, her dark eyes sparkling, her lips pressed firmly together in a sweet smile that hides a gap where the tooth fairy made off with her due. Andrea has been the idol of her little brothers; a treasure to her parents. Miriam floats about the house preparing supper and comes out from the kitchen often to watch her boys, regarding them tenderly through big, sad eyes. She hugs them, laughs with them, coos in her soft lyric voice. The door opens and Milton walks in. "Papi, papi," shout Aaron and Mauricio. They leap into his arms.

Outsiders say that the Flores family has always been close, but they are especially so these days. Two months earlier, on July 25, 1992, they visited the depths of despair. Andrea was playing, much like her brothers, and accidentally swallowed a small piece of plastic. It caught in her throat, and she panicked. She ran to her parents to tell them her trouble, then slumped to the floor. Miriam cried out. Milton wrapped his arms around his little girl and forced her diaphragm, but nothing came forth. Within minutes they were at the public hospital. But within seconds of falling to the floor, Andrea had left them. The doctors say that she died of a heart attack. Miriam and Milton were overcome. There, in the public hospital, with poor folk around, waiting their turn, they drowned in grief, in shock, in helplessness.

"Many people sent books and said things, but nothing anyone else could say could help me," says Milton. "God had to speak to me Himself." Miriam found comfort only in reading the Bible. Yet, friends and family brought strength through solidarity. "What helped was seeing how many people showed up at church to grieve with us," says Milton. During the entire funeral, he could remember but one condolence. An old uncle came all the way from Choluteca to the service. He gave Milton a hug and said, "I have come here to weep with you." When they arrived at the cemetery on the outskirts of the capital and stood by Andrea's grave, marked by a bed of flowers, Milton turned to Miriam and said, "My uncle has come here to weep with us."

His uncle's words held wisdom: To alleviate burdens, first weep with the suffering. In El Jute, for example, the fear of helplessness runs deep. When a campesino catches a fever, he worries that it might be mortal. When a child's stomach cramps, the pain could be passing or fatal. Milton visited El Jute often after Andrea's death. He tilled and planted in the field with Roberto, Benito, and the other campesinos. They helped him with their quiet presence. They helped by confiding to him their own pain at losing children. Still, Milton wept. One day, the woman in the neat turquoise house admonished him to dry his tears. "It is no good to cry too much for your little girl, for she is an angel," she said. "Angels become dirty with your tears." This the farmer's wife knew, for a woman in the village had lost her little girl of six and had become so desperate that for months she went each day into the forest and cried and cried. Then one day she saw her little girl, and she was all dirty. She saw what her tears had done, and after that she cried no more: Human tears can only soil one already perfected.

Shortly afterward, Milton had his own vision while away from Honduras at a conference in Panama. He awoke from his sleep with the strong sensation that he had been weeping for Andrea, and she had been standing there, watching him. "Daddy," she cried out, "how is it possible that I am causing you such a terrible pain?" Andrea began to cry herself, and Milton wondered, "How can someone we love so much make us so bitter?" He felt desperate, impotent. Sometimes the desperation boiled over into anger at the ones he loved most, so family feelings cascaded. Andrea had been Aaron's best friend, and he copied her every move. Now Aaron's playmate was his little brother Mauricio, who imitated him. He did not always enjoy it.

The campesinos of El Jute felt more loved than ever, because Milton shared his grief with them. It was sincere; it was an honor. "These guys must really care for us," Milton overheard one say to another. And Milton, in his grief, saw that he loved his family more than ever. "After God, my family is the only thing that is important. I didn't know that before my little girl died," he said. Indeed, in another period of reflection, God brought a verse from the Bible to Milton's attention, which he considered as much of an answer to the mystery of Andrea's death as he could receive in life: "do not be surprised at the painful trial you are suffering, as though something strange were happening to you. But rejoice, that you participate in the sufferings of Christ, so that you may be overjoyed when his glory is revealed. . . . so then, those who suffer according to God's will should commit themselves to their faithful Creator and continue to do good" (Peter 4:12–19 New International Version). So Milton and Miriam resumed their routines at work, home, and church—all routines, that is, but one: They no longer felt like playing music.

The Floreses were not the only ones among the cast of leaders in human farming to experience personal storms. But their revelation that family came before vocation—even before helping others—was significant. Don Elías, despite his focus on the campesino family's holistic development, found by the late 1970s that ties to his own family in Tegucigalpa had frayed. Liliana was not happy with his earthy calling, and Elías was not happy without it. By 1980, he had chosen vocation over marriage. He and Liliana separated and divorced. Their three children, José Elías, Liliana and Mireya del Carmen, went on to excel in their studies, displaying the independence of their father and polish of their mother, but the family unit as they had known it ceased to be. It was not a subject Elías liked to discuss.[1]

Candida, too, found her family rent apart. After she began working at Loma Linda, the man in her life, a taxi driver and the father of her three children, would come to the farm each day at the end of his shift and drive her back to Tegucigalpa. Some mornings, Candida would arrive at work black and blue. Friends said that her man drank; he was jealous. Yet his mother, who looked after and raised their children while the parents worked, was at times more mother to them than Candida herself. It was not a unique predicament for a working-class Latina, and it was not a situation that Candida expected to change. Nevertheless, after years of cooking for campesinos at Loma Linda, watching their confidence grow, their attitudes change, Candida decided to transform her own life. She would leave Tegucigalpa, Loma Linda, her family, her children, her country, and immigrate to Florida to join relatives. She would begin anew.

And so Candida left Honduras in 1991 and set out on her own. To his surprise, Elías felt himself suddenly on his own too, despite the fact that he was rarely alone. He hired other women to cook and manage the farm's services, but he found no one with Candida's commitment to the farm. "She was not an employee," he said to himself, "she was an honor." He suffered another loss as a rift developed in his relationship with World Neighbors, whose legal status in Honduras depended on ACORDE's sponsorship.[2] Through all the turbulence, Elías missed Candida's calm, consistent assistance. He missed Candida.

By that time, Fernando and Vilma had weathered their own times of turbulence. As the Tierra Nueva project drew to a close in 1989, the Andrades had a choice to make—to settle in Minas de Oro or return to Linaca. Fernando saw great potential in the land around Minas de Oro. The region had good soil, and with his savings from the project, he could buy a nice farm. His experience with Tierra Nueva had been positive. He had become friends with the American couple for whom he worked, and they had even taken him on a trip to the Pacific Northwest, where he visited the project's donors. Vilma, however, wanted to go home, and home was Linaca. Their eldest daughter, Nora Cristina, was there with her new husband, and so was Vilma's family land. Moreover, the family would be close to hospitals in Tegucigalpa in case anyone fell ill.

The truth was that Vilma had never felt included by the expatriates at Minas de Oro. She knew that they had had an important influence on her husband, but she had never felt their warmth. For two days, Fernando and Vilma argued about their future, and finally Vilma gave an ultimatum. "You can buy land if you want," she told Fernando, "but I am going back to Linaca." Fernando held his tongue. In the past he might have ordered everyone to stay, but Vilma knew that he had changed. She had watched him grow in his faith, and she knew that his actions would follow. Fernando supported Vilma's decision, packed up his life in Minas de Oro, and transplanted himself and his people back to Linaca.

With the money he had saved he bought his family a new tile floor and planted a large patch of garlic as a cash crop. He was able to pay for Nora Cristina to go to a private school to learn accounting. He still could not afford a special school for Sonia, who had a learning disability that public schools lacked the resources to handle. But Sonia, who had been at home helping her mother since finishing the sixth grade, began to receive a different kind of education from foreign visitors who came to their house to see the Andrades' corn.

The return home to Linaca was far from triumphant. Some neighbors continued to resent Fernando's success. Most disappointing was the fact that his brother, Edilberto, had given up conserving and improving his own soil. Edilberto's corn, across the road from Fernando's, was only six feet tall compared with Fernando's ten-foot stalks, and garbage was strewn throughout Edilberto's yard. Fernando's brother had taken a salaried job as a rural extensionist in a government project and, ironically, had no more time for conserving and enhancing the soil on his own property.

One of Edilberto's friends, a police administrator wide of girth and long of speech, has recently come back from a visit to Taiwan sponsored by the Justice Department. Paying a call on a rainy day, he is full of pessimism for his own country. He says that he is convinced that grassroots steps, a campesino improving his land and the well-being of his family, will never amount to more than a drop in the ocean of Honduras's development needs. "Honduras is hopeless," he proclaims. "Development is impossible without a scientific administration."

Honduras lacks centrally planned industry, he says, and it lacks a nerve center of know-how. Just consider the intellectual corps that Chiang Kai-shek brought to Taiwan from China when he fled Mao Tse-tung's forces. Just consider the phenomenal wealth Taiwan has created and its successful agrarian reform. "But what about the costs of Taiwan's economic growth?" asks one of his listeners. "What about the dead rivers, polluted cities?" This could be cleaned up, according to the police administrator. But Honduras's future? That is a dead end. Campesinos might as well give up. Even if they could produce more, they would have no market, and the government cannot buy excess production. Indeed, Edilberto has given up. Yet behind the rain pelting down that day, another storm is brewing, and this time it will blow not against campesinos but against the very central powers in whom men like the police officer thought it best to place their confidence.

Notes

1. Liliana Sánchez declined an interview for this book and requested that Elías serve as the source for their family history.
2. By the late 1980s, a split had begun to occur between World Neighbors and ACORDE. In 1989, World Neighbors' development activities suffered from a problematic expansion into store-based marketing of agricultural products in Tegucigalpa, which strained relations between World Neighbors and ACORDE. Although the idea of marketing agricultural products had been a dream of Elías's, he felt inadequately consulted as ACORDE

director and World Neighbors/Honduras' legal sponsor on the decision to vertically integrate and expand programs. In addition, he took exception to World Neighbors' failure to train and move Honduran staff into key management positions held by expatriates and over what he perceived as an emphasis at the time on technology transfer. Sadly, the mistrust and pride that can undermine community efforts also can be a stumbling block within the organizations that seek to assist communities. The leaders of neither World Neighbors/Honduras nor Loma Linda were immune, and the disease cost them dearly in emotion, time, and energy. In 1991, after a series of negotiations, World Neighbors legally separated from ACORDE and became an independent international NGO with legal status in Honduras.

8

Hands for the Harvest

ON HONDURAN mountains and hillsides, empowerment of the campesino had been radiating from Granja Loma Linda for a decade. By 1992, the human farming movement had spawned more than a dozen independent training farms across the country. What's more, government and donor organizations alike had accepted the tenets of human farming as foundational to successful rural development on any scale. From assistance to individual farmers and villages to schemes for regions and the nation, experts were referencing Loma Linda and its techniques. And the hillside farmers were not alone in finding their voice. From the low country and jungles, the forests and riverbanks, pockets of marginalized Hondurans, subsistence farmers, hunters, and gatherers, had begun to move, to stand up, to lay claim to their rights and resources. Loma Linda was just one important current in a tide of change sweeping the country.

> *In the end, people will have to associate their individual success with community success.*
>
> —LORAL PATCHEN
> U.S. Peace Corps
> volunteer

Replication of Don Elías's approach had taken place in corners of the country as remote as the southwest department of Intibucá, near El Salvador, where a seed planted at Loma Linda in 1980 bloomed a full decade later in a little village called Semane. One caretaker of the seed was a thin, serious-minded young Peace Corps volunteer with an unruly cascade of dirty-blond curls. Loral Patchen had transplanted her small-town roots in Rockton, Illinois, to the Ivy League before heading south to a Central American village. When she joined the U.S. Peace Corps after graduating from Brown University, she was determined to learn from local Hondurans and to promote development that honored all the bywords of excellence: grassroots, people-centered, transformational. She wanted a posting that called for her to live with the poor, as the poor. Semane, with neither phones, electricity, cars nor regular bus service, fit the bill perfectly.

Indeed, this border region was so cut off from the trappings of government that until recent years, people had used El Salvadoran colons as currency. After Loral completed her Peace Corps training in soil conservation, she lurched across mountain tracks with her bundle of belongings to the tiny campesino town and discovered another piece of grassroots authenticity: She would have to build her own house. Meanwhile, she could stay with a local farmer and his wife, Gregorio and Candida Velásquez. And so her learning began.

Although few inhabitants of Semane could read, Loral soon learned that they liked to quote, and one of their favorite bards was a man named Don Elías Sánchez. If the topic was soil conservation, their response was inevitably a pithy phrase from the master and servant of Loma Linda. "Licenciado Elías Sánchez always says . . . " became a familiar refrain. "They quote him endlessly," thought Loral. "In some ways it is like a cult, but in others, it is just a voice that they hear." What they heard, she recognized, was belief in their own success. Elías had convinced them that they were gifted and could change their lot, and that was the key to long-lasting development. "It's not Elías who has become deified," she concluded, "it's the ideas."

Loral also learned that Don Elías's voice had taken years to penetrate through the mountain mists of Intibucá to her new home. Her host Gregorio was ultimately the conduit. A decade earlier, when Don Elías first began to invite campesinos to Loma Linda, Gregorio had been part of a group sent by Save the Children to Loma Linda as part of a project to increase water resources for home use and family gardening. The campesinos had seen the fruits of human farming, but the enduring fruits of the project were water technologies: One-quarter of the eighty families in Semane had developed running water systems with cisterns and hoses. The added resource aided household chores and hygiene but did little to address an increasing health problem: malnutrition. Erosion of Semane's sandy soil was robbing farm yields, depriving children of needed food staples, and threatening to push the families from their land.

Few families could cultivate enough to feed themselves, and most farmers had to hire themselves out to coffee plantations to pick beans for a dollar or two a day. Four months of the year, many families had little more than tortillas and salt to eat. Thus had survived Gregorio, Candida, and their eight children and one grandchild. By 1988, however, the Velásquez family was ready to give up, ready to find another mountainside to slash and burn and plant corn on. That same year, the U.S. Peace Corps brought assistance to Semane. It invited a group of farmers, including Gregorio, back to Loma Linda for training, and when the farmers returned to their village, the Peace Corps followed up.

By the time Loral arrived, Gregorio's farm had become a model for the community, one of Loma Linda's many spontaneous satellites replete with human farming technologies and increased corn yields. Gregorio's success meant that he could afford to stop working the coffee plantations. Instead, he opened his farm once a week to train others to love and care for their land. If visitors were skeptical at first, they were cured by the sight of ten-foot stalks of corn dwarfing Gregorio's five-foot-three-inch frame. His crop stood more than double the height of local maize in the best of years.

Loral worked to build her mud-brick house with help from her neighbors while lodging at Gregorio's. Candida Velásquez, a woman who had never seen a paved road, became the best friend the twenty-four-year-old college graduate had ever had. Loral observed that the farmers in Semane who were willing to take risks usually had very supportive wives. Tiny, round Candida, her warm smile stretched around one front tooth, had battle credits to prove her fealty. She had delivered five of her children herself, feeding the others between contractions as Gregorio carried on his work in the fields to provide for them. Candida watched her husband's relationship to the family change as the land responded to his care. Two of his sons, seventeen-year old Alejandro and eleven-year-old Mario, were following in their father's farming footsteps. "Our home has become happier," she confided to Loral.

Gregorio, who had attended three months of literacy training, volunteered to lead Bible readings for Semane's Catholic community as part of the local lay movement. He studied the Bible in the evenings by kitchen candlelight, carefully learning the texts he would read, and by the end of 1991, he was leading the celebration service three weekends a month. That year the Velásquez family set a community precedent: They hosted the Christmas worship meeting in their home. Gregorio read, and Candida made bread and coffee and provided tamales for seventy guests. In so sharing, they brought full circle the wealth they had achieved. Loral, raised in a family of atheists, was eighteen before she entered a church. But she grew to love the services in Semane. "People wonder why so many campesinos are religious," she mused. Yet, it was obvious to her. "It's impossible to deny the divine when it's all around you. You walk out your door, and you're in Creation." She too saw the signs.

She also saw discord. The greatest farming challenge in Semane was to grow a caring community. Although the villagers were brothers, sisters, cousins, kin, they were not supportive family. Houses were spaced far apart, jealousy and distrust reigned, and alcoholism thrived. Apart from the Delegates of the Word lay movement, no apparent tradition existed around which to build common goals. Indeed, many people who acquired knowledge of soil conservation hid their know-how. "They don't want others to succeed," observed Loral. How long would it be before they realized that community success was the best guarantee for their own success? How could she help them? One day, Gregorio suggested to the lay community that they build a new church meeting hall, and his listeners were shocked. "Little by little we can do it," he said. He was taking principles he had learned at Loma Linda and applying them to life: starting small, achieving with excellence, sharing what he had learned, creating personal and communal satisfaction, and using God's wisdom. Soon, other villagers began to talk of building a new school to supplement the current one, which only went up to sixth grade. Gregorio dreamed of a secondary school for his children, and he realized that only a common front could win over government authorities and bring a high school to Semane.

Loral saw other problems: Although many Semane farmers built terraces, contour ditches, and live barriers to *conserve* their soil, few took steps to *improve* it. The local supply of organic material was limited, and compost could only go so far. After she moved into her own little bungalow, she hurried to plant a garden with corn and velvet beans and demonstrate the effects of cover-cropping. The velvet beans crawled over everything, around and up the corn stalks. Such chaos frightened the farmers, whose fathers and grandfathers had taught them that good corn came from "clean" fields—no weeds, no stones, no vegetation. Loral had learned of CIDICCO in Peace Corps training and called Milton Flores, who rumbled over the mountains with his slide show and battery-operated projector to make a presentation. When the local teachers heard that there would be a spectacle, they sent the sixth-grade class to join the farmers and unwittingly primed Semane's next generation of agriculturists. Milton toured Gregorio's farm and invited a group of farmers to El Jute to work for a week with Roberto Zepeda. When they returned, they were quoting Milton, Roberto, and Gregorio, and they were ready to risk change.

Similar stories played out around other Loma Linda alumni. Sometimes the seeds planted by Elías took years to germinate, sometimes only weeks or months. Florentino Santos Villeda of La

Crucita (a village near Socorro in the department of Comayagua) was a fast bloomer. Within three months of visiting Loma Linda with World Neighbors, he had built up his own farm to serve as a model for his community. Above all, he was motivated by the low cost of Loma Linda's innovation. "I saw that he had done most of [his work] without money," said Florentino, "and I had everything I needed to do the same things on my own farm." Indeed, he had three generations of heads, hands, and hearts. The Santos's 6-manzana (10-acre) farm had passed from father to eldest son since the time of Florentino's great-grandfather. Today, grandfather Adan, seventy-four, still works the farm with Florentino, forty-one, and fourteen-year-old Isai, the eldest of Florentino's five children.

Florentino disliked Elías's gruff manner, but he appreciated what he saw and believed what he heard. "He was direct and heavy with everyone," said Florentino. "He just shook our hands with a straight face . . . and started talking about hygiene. He asked the men how many times they changed their underwear . . . then he asked one man to describe a pig . . . so we saw that he was trying to make the link between pigs and our hygiene. And then he asked how often we bathed. One man said four times a week, and each would say a different amount . . . but one man said nothing, so Elías said 'and when do you bathe?' and the man admitted with embarrassment 'only in the dry season.' We were expecting a much gentler treatment and much more courtesy . . . I was offended, but I found out later he thought we were agronomists."

Florentino also found himself inspired. He knew when he saw Loma Linda that he could transform his farm, and he believed, as Elías had emphasized, that good farming was a physical and spiritual pursuit. As Loral and others had observed, even when Elías failed to click with someone, when he got caught in his own double standard for campesinos and professionals, his program seemed to work. He could be painfully direct, but the truth of the method and sincerity of his commitment prevailed. In the fall of 1992, World Neighbors/Honduras held its annual meeting in Socorro, Comayagua, and featured a visit to the Santos farm. At the foot of his hillside, Florentino made a wooden welcome sign that pointed to a path of newly built steps snaking up from his old fields of cane, banana, and corn along an acre of masterfully cut terraces. A World Neighbors extensionist has come by each Tuesday for the past forty days to help him transform his land.

The physical transformation began with building compost piles, then carving terraces and tilling and enriching strips of soil. Florentino was an artist who worked with nature's palette. He made the most of Creation's gifts and devised ways to live with its demands. He went on to plant corn and cover crops and dig contour ditches every few meters to catch rainwater and force it to seep into the soil beneath his plants. He planted tomatoes above the corn, where they would catch the last shafts of daylight, and soy to serve as an ingredient for nutritious recipes his wife had learned from World Neighbors. He sowed marigolds, whose scent repelled insects, amid his rows of vegetables. And he planted strips of sorghum, whose tufts of seeds grew above the corn and drew birds away from staple crops. After years of trying to keep the predators off his farm, he had devised a way to live with them. Around his home he built a sturdy fence and beside it planted flowers, the smiles of the soil.

"My farm takes much more time now," admitted Florentino, "but because it is productive, I can spend more time at home." He has already reduced the days per week that he hires himself out as a construction hand. Soon he might afford school shoes for his children, who gather barefoot at the fence gate to wave good-bye to their visitors. One visitor, Milton Flores, accidentally enacted a parable of community development on his departure.

As he left the Santos farm, Milton's pickup became mired in mud. It had been raining all week, and rivulets of water still flowed down the steep dirt track. He jumped out of the cab, wedged rocks underneath his tires, hopped back in, and gunned the motor. But his wheels continued to spin. A crowd of children ran from their homes to watch the spectacle, and farmers began to make their way from the fields to offer assistance. The first three men to arrive pushed with all their strength as Milton again pressed the accelerator.

The pickup shifted forward a few inches, then fell back deeper into the mud. Another man joined the three farmers and they strained together, but the pickup sank again. They tried once more, stooping to place more rocks under Milton's tires, and when another farmer arrived, together the five threw their weight against metal and gravity. The tires spit out mud and debris, then caught and carried the pickup forward. A loud cheer arose from the farmers, and they jumped into the truck bed, riding with Milton to the crest of the hill. By working together, learning from failure, and experimenting with new solutions, everyone had moved forward. If Florentino's farm could inspire others to improve their land and conditions, the community as a whole could advance. But the first ones to join the bandwagon would have to push.

Some of the people Elías inspires live not on hills but in the city, bringing the principles of Loma Linda to urban development. A case in point is Ramón Romero, a philosophy professor at Honduras Autonomous University who has become an urban proponent of human farming. Equipped with a doctorate in political philosophy and ethics from Florida State University, he was exploring the tensions between development and ecology when he decided that he should get practical and garden. Ramón drove to Loma Linda to seek instruction on vegetable gardening, and Elías promptly invited him to join a class of campesinos. Ramón found himself in the most unique and practical philosophy course he had yet to encounter. The concept of human farming meshed well with his emerging view that as society finds itself in crisis, the most "civilized" people are not those who exploit the most but those who care the most for Creation. "Human growth has been expressed in terms of possession," says Ramón. "He who has the most, is the most. But the human farm challenges this proposition, because it expresses human growth in terms of transformation, developing human capabilities and collective potential: He who becomes the most relationally, is the most."

Ramón left Loma Linda with a promise to himself to do something, however small, to further the kind of development work he had encountered at the farm. His commitment was contagious:

As soon as he returned to his household, his family joined the work in earnest. Ramón's wife, an attorney, went off to the market to scavenge vegetable peelings and corn husks for a compost heap, and his five children set to work transforming forty old tires brought from car repair shops into Loma Linda–style planters. Their house, which hung together like Lego blocks—a room added here, an entire floor there—had only a small yard but a large roof. On top, up the stairs past Ramón's crowded study and out under the stars, would sprout their urban garden, their human farm.

Within weeks, the Romeros were growing carrots and broccoli, cauliflower and sweet peppers, and had planted the garden with corn. "This activity has become more and more important to us," says Ramón, "It's a therapy for stress; an activity for the family; part of our promise to society and creation; and we can eat the vegetables without fear of contaminants." Although their neighbors have not rushed to follow suit, the Romeros have opened their home—and their roof—to urban youth groups interested in the movement. Many of Ramón's rooftop discoveries appear in his lectures and writing. Like his teacher Elías, he plans to persist and insist on his message.

For government and donor organizations, such perseverance was as necessary for the office-bound agents of development as for the farmers in the field and town. In the late 1980s, the United Nations Food and Agriculture Organization (FAO), a major international donor, began to doubt the impact of its food aid and technical programs in Latin America and turned to promoting farmer development. Carlos Zelaya, FAO's representative in Honduras, ordered up a crash course at Loma Linda and two other training farms and sent 400 campesino community leaders for two days of technical instruction and inspiration. Afterward, the leaders returned to train their communities, with assistance from the Ministry of Natural Resources. "It's a relatively new approach for the FAO in Latin America and the Caribbean," says Zelaya, whose office shelves are crowded with technical reports and binders full of data. "We need to gain the trust of people, and we're not going to do it via statistics." When questioned on results of the new program,

Zelaya pulls out not a report but two small photos taken two years apart. In one, a farmer from the FAO-sponsored program is standing amid weeds next to old fence posts; in the second, the same farmer, in the same T-shirt and in the same spot, is dwarfed by a field of corn.

Where the FAO seeks evidence of impact on a national level, Zelaya points to improved nutrition among many campesinos. "This is not isolated, this is widespread." Specifically, he says that the FAO has gathered evidence that improved nutrition is improving children's ability to fight intestinal worms in the program zone. "It's like working in a petri dish," he says. "We're creating bacteria for a positive infection."

One person so infected was Victor Inocencio Peralta, whom Zelaya calls the Pancho Villa of Elías's home province of Choluteca.[1] Victor, one of the region's most powerful campesino leaders, was a strong-arm type who wore his sombrero with pride. In fact, he refused to take it off for anyone—not in a home and not even, he maintained, for the president. In his hat he was a man. Elías's first move when he met Victor was to welcome him and take his sombrero. "That," says Zelaya, "is the beauty of Elías's technique. He takes for granted that people want to take their hats off, that they want to bathe. He inspires people to do things by assuming they want to." At the end of the course, Victor was down off his high horse and returned to work. "He was a great leader," says Zelaya, "but he decided to be a different kind of leader—one who produces."

Zelaya says that the FAO has seen something else in the petri dish: a worldwide multiplication of the germs that propagate human farming. "There are Elías Sánchezes emerging all over Latin America. They're emerging as a natural response to the problems of agriculture," says Zelaya. "And it's not just a Latin American phenomenon—there are Elíases throughout Africa and Asia. They are trainers who have prepared themselves through study, then bring to the field three essentials: thought, action, and commitment."

Even the U.S. Agency for International Development (USAID), the largest bilateral donor to Honduran development, was trying to reflect these essentials in its 1990s approach to land conservation. USAID's Land Use Productivity Enhancement (LUPE) project, scheduled to operate from 1990 to 1998, hoped to reach 50,000 families through the creation of about ninety local extension service agencies that would stimulate change by working with individual farmers. The strategy offered a departure from USAID's historical approach, which had focused on the diffusion of new technologies through the Honduran Ministry of Natural Resources. There was much evidence that change so engendered was unsustainable; when programs withdrew inputs, farm innovations often ceased.

In 1982, for example, USAID launched its first comprehensive, campesino-based soil conservation project, a precursor to LUPE. The Proyecto de Manejo de la Cuenca (the Watershed Management Project) provided hillside farmers food-for-work subsidies to introduce soil conservation technologies on their properties. The project, which aimed to stop erosion, emphasized physical structures such as rock walls and ditches. Observed a USAID contractor familiar with the project: "Once the subsidies were withdrawn, the structures tended to decline pretty rapidly."

In 1989, in an attempt to learn from its mistake, USAID funded the LUPE project, an eight-year, $50 million successor to the Manejo scheme. LUPE covered a larger area and added farm productivity enhancement techniques such as green manures, composting, natural pest repellents, and home gardens to soil conservation technologies. It did not offer subsidies to farmers who participated in training and adopted suggested agricultural methods; it offered only the promise of better yields.

Sadly, however, problems engendered by LUPE's failed predecessor were enormous: Twenty-six of LUPE's local extension agencies were inherited from the Manejo project. Summed up the contractor: "Unfortunately the food for work [in the Manejo project] contaminated not only the farmers but also the extensionists with a very unproductive mentality. . . . The extensionists got used to not having to sell technologies. They got used to playing Santa Claus. A lot of these Manejo folk still work for

LUPE. They're good people, but it's taken them a long time to throw off the old mentality."

The LUPE directors were beginning to realize that the most important aspect of their work was to convince farmers of their *need* to change, to sell people on their own abilities, not just technologies. Indeed, some say that Elías itched to take over the program and transform it, Dale Carnegie style, into an exercise in inspiration. For now, LUPE was trying to overcome past precedents, and the jury is still out on its chances for success.

But the jury was in on the ability of change to trickle up from the grassroots, both from Loma Linda and from other centers of energy and daring. By 1992, the NGO community in Honduras had exploded on the scene, lobbying for and winning major policy battles with government. From the door of Milton Flores's peaceful office hung a hand grenade that helped trigger a countrywide explosion—a small pinecone. The artifact represents the power of an unprecedented movement fueled by ecology-minded groups that reached a crescendo in February 1992. That winter, thousands of marching, pinecone-carrying Hondurans managed to send packing a multinational timber harvester that had contracted a forty-year concession with the Honduran government to log the nation's tropical pine forest reserves of La Mosquitia.[2] It was the first such victory for the conservation community, and the story played out in the corridors of government, in the streets, and on the airwaves. Its unfolding demonstrated that peasant farmers, Indians, environmentalists, and other marginalized groups had learned to wield a powerful new weapon: public opinion.

Some contend that the tale of Stone Container Corporation in Honduras begins at the turn of the century with patterns of direct foreign investment and old-boy deals established by banana magnates like Sam Zemurray.[3] But our abridged story begins with an item in the Tegucigalpa daily *La Tribuna* on September 25, 1991: "The government of Honduras and the North American company Stone Container have made an agreement to develop one million hectares [2.47 million acres] of forest in the departments of Gracias a Dios, Olancho and Yoro to produce and export wood chips to the United States and other countries." The article went on to mention that the activity would create more than 3,000 blue-collar jobs and could double Honduras's foreign exchange earnings from forest products in three short years, while improving the forests themselves through a program of selective harvesting, reforestation, and afforestation. Above all, it said that Chicago-based Stone Container would not compete with Honduran forest product firms but rather salvage residue and create a market for pulp. Hondurans just didn't buy it. What's more, they let it be known.

The national College of Professional Foresters opened fire, declaring the accord illegal. In a letter to *La Tribuna* (October 7, 1991), the foresters inferred high-level horse trading of natural resources in a manner long abolished under laws of the Honduran Forest Development Corporation (COHDEFOR). The Mosquitia community, indigenous peoples inhabiting the coastal pine forests of Gracias a Dios and Olancho, shot back. They saw economic benefits in the action and wanted their stake. In a half-page pronouncement in rival daily *El Tiempo* (October 22, 1991), the people of La Mosquitia affirmed the government's decision and solicited a role in negotiating and executing the contract and supervising the activity.

By the end of October, national and international environmentalists were up in arms denouncing the accord, with the media right behind. Stone Container scrambled to improve public relations. In a front-page article in the English-language weekly *Honduras This Week* (November 2, 1991), company spokespeople enumerated environmental precautions to be taken: no clear-cutting, no heavy equipment or large crews, no paved or gravel roads. But they acknowledged that the company had not conducted an environmental impact study of the zone. Stone Container invited a pool of reporters and ecologists to visit the site to be logged; it even whisked some of them off to its operations in the United States, where the firm contracted with private landowners to manage their woods in return for rights to harvest their timber. The company said that it wanted to do the same thing in Honduras, to "give the local people a chance to help harvest the forest" and create their own wealth.[4]

But the media were unconvinced, and a local radio announcer called on protesters to carry "peace grenades," small pinecones. Practically overnight, thousands of citizens began bearing cones on their vehicles or persons. A scathing *El Tiempo* editorial accused the government and Stone Container of duping Indians and campesinos into thinking that they could win easy money by trading off their natural resources, their land, and their subsistence.[5] For that matter, asked *El Tiempo*, did the indigenous people of La Mosquitia actually have rights over the forest? Weren't the forests, in fact, a public good?

Questions of who was manipulating whom in the name of "development" and what were the best interests and rights of Honduran campesinos, and Indians, flora, and fauna moved from the back burner to the dinner table of average Tegucigalpan households. A national debate began over groups and regions that were often out of sight and out of mind. By the time USAID brought a new environmental adviser on board in February 1992, public protests against Stone Container had reached a fevered pitch. "It was on the front page of all four local newspapers every day," says a North Carolina forester named Margaret Harritt, who took up her USAID post on February 17, 1992. "There were demonstrations at least weekly, sometimes twice a week. They blocked off the streets. There was amazing public pressure."[6]

The anger boiled over into graffiti, marches, and public speeches. The people of La Mosquitia wanted partnership; the environmentalists wanted preservation; the foresters wanted due process; the taxpayers wanted clean politics; and, in the end, the deal makers wanted out. On February 27, the government announced that negotiations with Stone Container were suspended for "technical reasons" and "in the interests of the public."[7] Environmentalists declared victory, and the hunters, gatherers, farmers, and fishermen of La Mosquitia went to work with a heightened profile and resolve to prove their rights to the forests in preparation for future battles. By their side was a man well experienced in empowering people and communities: Don Elías Sánchez's former partner, Wilmer Dagen.

Notes

1. Pancho Villa (1878–1923), originally named Doroteo Arango, was a Mexican bandit and revolutionary who became a folk hero as both a "Robin Hood" and fighter for social reform.
2. La Mosquitia, depicted by Hollywood and actor Harrison Ford in the film *Mosquito Coast*, is a traditional name for a section of the Atlantic coast stretching across Honduras and Nicaragua and inhabited by indigenous peoples, both Indians and *ladinos* of Spanish descent, who hunt, gather, fish, and farm. In Honduras, La Mosquitia includes the departments of Gracias a Dios, Colón, and Olancho, and has a population of about 40,000. It is widely considered the region of Honduras least privileged by government services and official development efforts. MOPAWI Biannual Report, 1990–91 (Tegucigalpa: CADERH), p. 3.
3. Editorial, "La contrata con la Stone Container y su no aprobación en el congreso nacional," *El Tiempo* (Tegucigalpa), Dec. 26, 1991, p. 6.
4. "Stone says project will benefit people, environment," *Honduras This Week* (Tegucigalpa), Nov. 2, 1991, p. 1. Interview with Mark Lindley, manager of corporate affairs, Stone Container Corp., Chicago, June 28, 1994.
5. Editorial, "La contrata con la Stone Container," *El Tiempo* (Tegucigalpa), Dec. 26, 1991, p. 6.
6. Interview with Margaret Harritt, environmental adviser, USAID/Honduras, Tegucigalpa, Sept. 30, 1992.
7. "Stone agreement canceled," *Honduras This Week* (Tegucigalpa), Feb. 29, 1992, p. 1. In an interview in June 1994, Stone Container Corp. spokesperson Mark Lindley said the firm had no plans to go back to La Mosquitia, but still thought it was an area of Honduras that held great promise. "The program makes so much sense for that part of the country . . . there is so little employment and people are slashing and burning."

9

A Human Farm

OUTSIDE THE Andrade's Linaca home, the air is heavy and skies are dark with thunderheads. Inside, Sonia is cleaning the kitchen's huge wooden dry sink, and Juanita, who has finished her homework, looks after Nora Cristina's baby, Jili Michele. It's a typical Sunday, and burdens seem light as family members go about their chores. The Andrades have been back from Minas de Oro for over two years, and Fernando has improved on his Linaca technologies. He is particularly proud of his latest brainchild, a raised wooden box filled with soil that serves as a worm breeder for the vegetable garden. Vilma's mind is on her latest commercial project. With proceeds from her garlic patch, she saved enough to buy a little heifer, and now her cow has calved. Already the family is getting fresh milk each day, and Vilma is saving to buy another cow to increase her production and sell cream and cheese.

The living room has changed greatly from the days when

International organizations should act as yeast—they should help national efforts rise. . . . Spark plugs are the only specialist we need in development.

—DON JOSE ELIAS SANCHEZ

97

Fernando used to fight soil erosion by slashing and burning more forest. It is festive, hung with green and yellow crepe paper and a picture of an anonymous American bride. To one side sits a hand-made cardboard cathedral; at the room's center, a velveteen couch and chairs surround a color television set. Mounds of red and black beans sit neatly piled in corners on the new tile floor, ready for winter planting. Fernando's first love, however, remains his corn. Despite portents of a downpour, he will trek to his hillside to see its progress.

Down the gravel road and across a swollen stream, Fernando splashes along in his slicker and sombrero. A campesino woman hurries past him back to town, clutching her straw hat, and two children trot behind with umbrellas. It's the sort of day when people leave their fields early, but Fernando marches on. Up over the ridge, he pauses to look down the valley at the whitewashed adobe cottage where Vilma was born. Her family no longer lives there, and the only activity he sees is horses and oxen grazing beneath a tulip tree. From the ridge, a narrow rock path winds its way up a steep hillside to uncultivated pastureland and pine forest, which mark the beginning of the Andrade farm. Here, Fernando leaves the path and makes his way through a quarter mile of brush and trees, across a rushing river and its muddy slopes, and finally to the foot of his cultivated hillside. A visitor once asked him why he didn't plow his fields closer to the path. "You know," replied Fernando, "that is exactly what I asked my father." Now, as he gazes up at his terraces of corn and beans and onions, he feels the first drops of rain on his weathered hands.

A storm is surely coming and nearly upon him. Fernando begins to climb up the steep slopes toward a toolshed and its promise of shelter. He crawls over a dry stone wall, past piles of soaked compost made from cover-crop residues and fallen leaves. The ground becomes slick; Fernando slips, then regains his balance and reaches the little hut halfway up the slope. Rain pelts down, pounding against the hut's tin roof. Lightning bursts across the sky, with a clap of thunder. It is not the first time Fernando has taken refuge in a storm, and as always, the wetness makes old wounds tingle: the scar on his leg, where he slipped with his machete while clearing fields; the place on his back where a log fell and bruised him; and the jagged line on his wrist, where he nearly lost a hand. Fernando has sacrificed for his land, and his land has given up its fruits in return; he has watered the soil with sweat and blood. But there is another wound, split wide on his return from Minas de Oro, that still festers.

The labors of human farming take enormous quantities of time. For a while, Edilberto Andrade enjoyed the sacrifice and reaped its rewards in both field and family alongside his older brother. When an opportunity for a salaried job in government extension with USAID's LUPE project came his way, he took it, perhaps viewing it as a chance to spread the knowledge he had gained. Whatever his motive, a steady slide in the health of his own farm accompanied his new occupation. Garbage accumulated in the yard, and corn yields shrank after he joined LUPE to train others. The problem may have stemmed from low morale at LUPE, which, despite high goals, faced an uphill battle to convince farmers used to receiving handouts to adopt new methods without gifts of seed or fertilizer. When Fernando discovered the state of Edilberto's farm on his return from Minas de Oro, he was deeply saddened. Surely, thought Fernando, LUPE's own extensionists should take time to conserve and improve their soil if they want to convince others to do so. Isn't that the essence of "good development"? Showing by doing? Managing your farm, your resources in a way that makes sense and produces more food for the family? Pride for the farmer? Honor for Creation? Fernando, in his hut in the rain, thinks of Edilberto's meager crop, and he hurts and hopes and loves his brother.

Fernando is convinced that love and trust reside at the heart of good development, fueling the ability to change. To love one's land, one's family, one's community; to be made confident through love to acquire knowledge; to be propelled by trust in one's knowledge to new action—yes, this is the secret to good development. Demonstrating such feeling in his sincere, imperfect way was Elías's greatest gift to Fernando. Now Fernando believes that, through God's grace, he has developed his own reservoir to care for others and think for himself.

As the rain lessens, Fernando makes his way home from the tool hut, sporting his latest technology—a plastic bag over his sombrero. He retraces his steps past Vilma's childhood home, past the little shed where he sleeps at night, and back to Vilma, his girls, and the yellow crepe paper that makes each day a fiesta. He wonders about the future of his land and hopes that Sonia or Juanita will marry a farmer. Maybe, he muses, his girls could learn to farm themselves. Then he shrugs: It will be their choice. He must trust.

At Loma Linda, Elías also recognizes his need to trust, to reach out in love and heal old wounds. With the legal separation of ACORDE and World Neighbors in 1991 and a changing of the guard at the latter (which put Miriam Dagen, Wilmer Dagen's wife, at the helm), relationships between Loma Linda and World Neighbors renewed themselves, and fruitful collaboration resumed. By 1992, each organization was functioning with autonomy and clearly contracting its roles in joint ventures. Indeed, transplanting World Neighbors from Elías's guardianship to independent legal status in Honduras constituted necessary organizational development. Each agency now had space to grow new shoots and to scatter more widely its respective seeds for change. Although the key to fostering good development is knowing when to nurture growth, when to transplant, and when to prune, such knowledge often comes only through hard experience.

So it was, too, that Candida Osorio's experience abroad helped her recognize a path to personal development in Honduras. After six months in Florida, she decided to renew old ties on new terms. To Elías's delight, she returned to her country and to the farm. She came not as an employee but as a partner in the enterprise and moved into a newly built set of rooms at Loma Linda. Slowly, surely, their professional relationship began to blossom into something deeper, something fragile that lifted a great sadness from Elías and gave him cause to soften.

The sadness at the Flores home was softening too, although the memory of Andrea remained strong. Aaron excelled at his studies, and little Mauricio, too young for school, learned to read on his own. Interest in CIDICCO continued to grow and foster change in Honduras and abroad, which kept Milton and a staff of four traveling, speaking, and planting seeds. Margaret Harritt, settled by the fall of 1992 into her new job at USAID, noted the flowering of NGOs from the handful that existed when ACORDE was founded to 400 two decades later, many combining missions to develop people with the enhancement of natural resources. The Honduran umbrella group for NGOs (FOPRIDEH) and the Honduran 1992 Environmental Agenda identified Granja Loma Linda and World Neighbors as trailblazers in agricultural development.

About this time, one of the most dramatic displays of grassroots development came out of the forests of La Mosquitia. A quiet force behind its evolution was Elías's former partner, Wilmer Dagen. By 1981, Wilmer and Miriam Dagen had served in the Philippines with World Neighbors for four years and returned to Honduras to move to its least favored region. With war rumbling across the border in Nicaragua, La Mosquitia was playing host not only to indigenous poor but also to thousands of Nicaraguan refugees. Wilmer began running a development program alongside relief operations for Wheaton, Illinois–based World Relief Corporation; Miriam, who had completed medical school in the Philippines, joined the rural health service. It became apparent to World Relief and to Wilmer that indigenous peoples of La Mosquitia were virtual refugees, unserved by official institutions and nearly voiceless in their country's politics. To reach them, they added a development component to their relief program for Nicaraguans, but then they went one better: Wilmer helped the people of La Mosquitia form their own NGO, with backing from World Relief and other donors and in partnership with a Miskito Indian named Osvaldo Munguia. Thus was born Mosquitia Pawisa (Development of La Mosquitia), or MOPAWI. Ten years later, with Stone Container fixing the country's eyes and ears on their region, MOPAWI's membership emerged to argue convincingly for indigenous peoples' rights to the land they hunted, fished, and farmed.

The unveiling of their evidence came seven months after the annulment of Stone Container's accord and weeks before the 500th anniversary of Indians discovering Christopher Columbus on their shores. In a two-day congress in Tegucigalpa, representatives of La Mosquitia's four indigenous groups—Miskitos, Pech/Paya, Garífuna, and Sumu-Tawahka—as well as native *ladinos* presented a painstakingly crafted land-use map to more than 200 politicians, bureaucrats, and private development organizations, showing ancestral names and uses of rivers, forests, hills, and dales. Indeed, land that appeared on the government's charts as "unexploited" was shown to serve a population that harvested renewable resources and had done so for centuries.

The Indians made use of the gathering to propose solutions to sensitive issues such as land tenure, socioeconomic development, conservation, and management of natural resources. And they took advantage of their high-level audience to demand greater social assistance and an end to human rights abuses by the military in their region.[1] For their part, key government departments, including the military, kept representatives in attendance throughout the meeting. The event, punctuated by Miskito songs, guitars, and the throb of tortoiseshell drums, marked a first major thrust by the shy natives of Honduran coastal forests to stand up and be counted. "It's probably the most transcendental event in La Mosquitia's development history, which dates back twenty or thirty years," said Andrew Leake, one of two outside experts who helped organize the map project.[2] "It's the first time that [these] . . . people have been really put on the map."[3]

The showing gave courage to other Honduran groups seeking a voice in decisions affecting their future. And it rekindled the interest of private and international institutions in helping Hondurans help themselves. MOPAWI, although operating in a separate sphere from Loma Linda and for a different cause, was sharing its learning and creating the kind of personal and communal satisfaction advocated in human farming.

Ten days after MOPAWI's presentation, one international institution, the Pan American Agricultural School, found its foot soldiers gathered around the red and white checkered plastic tablecloth of Granja Loma Linda's dining room. An instructor of rural development, Ernesto Palacios, had brought with him two Zamorano agronomists and fourteen campesino teenagers from villages reached by the school's rural extension program. After touring the farm and sharing stories and poems, Elías probes the youngsters' thinking.

"How do you identify people's needs?" he asks them.

On behalf of the group, Ernesto invokes several analytical models. Elías waits for another response.

One of the Zamoranos, Patricia Cruz, pipes up. "I met a family in one village, and I could see they had a nutrition problem even though they could farm well. The children needed milk and vitamins, so we brought them a goat."

Several of the campesino youths add that they have become interested in identifying their own farm needs through the work of NGOs in their communities.

"How then would you identify my needs here at Loma Linda?" asks Elías.

"Maybe you could have more animals," suggests one boy. "You depend on chicken manure to help your soil, but you don't produce enough on your farm. You are dependent on outsiders."

Another campesino, a young girl, asks, "What about saving water from your river for the dry season?"

Yet another asks, "Why don't you start a small farm store?"

The teenagers seem pleased with their ideas, and Elías, although strongly convinced that he has good reasons for purchasing rather than producing his chicken manure, bites his tongue in encouragement of the brainstorming. When the pupils' ideas are exhausted, the teacher asks them where all the information they have gained that day will go.

"To our brains," says a student.

"Where else?" asks Elías, then answers: "To your hearts." And he leaves them with a sobering thought: "Just remember, the world's development has a price. Calm, nature, safety. Contamination of your water, your ears, your eyes. We don't pay in money. We pay with our souls."

A teenager from the village next to Linaca volunteers that there are seven aid agencies, including LUPE and Zamorano, working in his town of 180 families, and the farmers have become savvy consumers. "The people just wait to see who will offer them the best package before they decide to participate in anything."

There is a lull. A campesino, a former trainee who has stopped by the farm to visit, clears his throat, stands, and speaks. "Don't worry about the chicken manure," he says. "You may have other garbage. Don't worry about the agencies. You have all that you need, if you will use it. "

With that, the session ends, and the two Zamoranos, Patricia Cruz and Laura Suazo, stop to chat with Elías. They are both twenty-five and have completed the school's rigorous agronomic engineering degree, with emphasis in rural development. Now they have joined Zamorano's rural development project as extensionists. "I joined because I wanted to be useful," says Patricia. "What I consider important is to start somewhere, and make sure you don't repeat others' efforts." Patricia and Laura come from the minority of Zamoranos who have chosen to work in the development sector, but it is a growing minority with voices that resonate. Senior managers at many NGOs—Catholic Relief Services, World Neighbors, CARE, CIDICCO—are Zamoranos.

What is the future of human farming beyond Loma Linda and Elías?[4] In truth, it will depend on more heads, hands, and hearts. Many have committed to the cause already, both in Honduras and beyond. Let Elías show you some pictures: There is Ismael Vargas, one of his original partners in developing Loma Linda. He's wearing a plaid shirt and white jeans and is bent over a row of sweet potatoes at his own training farm in Zopilotepe, Olancho. Next to him is a tall Peruvian in a red shirt, with hat in hand, who has come to observe human farming. There is Salomón now Mejía, who runs a training farm in Santa Rosa de Copán near the Guatemalan border. And here is Abel Ortiz, another former trainee, in a baseball cap and T-shirt, playing guitar among plump tomato plants and leaning against a well-sculpted terrace. There's a young Candida, first baking with a solar cooker, then washing dishes by the river, before Loma Linda had a kitchen. And there are groups of trainees from Lempira, Choluteca, Olancho, Comayagua, Copán, El

Paraíso, and Intibucá, posing with their hoes and wearing serious expressions. There is also a group of North American visitors, all smiles, with their cameras hanging from their necks. Last, one of Eliás's favorite pictures—two donkeys nose to tail, scratching each other's backs.

At Granja Loma Linda, Sunday is a day wholly dedicated to back-scratching, to relaxation and meditation—in short, to rest. With the Zamoranos gone, Elías and Candida can walk the farm, welcome friends, and listen to the cassettes of Charlie Pride music that Elías keeps from his days in New Mexico. This Sunday is a special one, because the Flores family will visit Loma Linda for the first time since Andrea's death. Candida, in her good blue dress, makes sure that the farm's mascots, two noisy green parrots, are in their tree ready to greet the children. Elías brings out his cassettes and his old tape player. Tires churn up the dirt track, a door slams, and gravel crunches under two little pairs of feet as Aaron and Mauricio race around the corner of the cookhouse to see the birds. "Welcome," says Elías, with eyes full of love. "Welcome," says Candida, now carrying a platter of fresh fruit bread. Milton, Miriam, Elías, and Candida embrace by the Rio Chiquito. They stand and talk and laugh. Beside them the children tumble about, and the river plays its music among the rocks, a song that can enter one's heart.

Notes

1. Eric Schwimmer, "Land use map presented in congress seeks to affirm Indian rights in the Mosquitia," *Honduras This Week* (Tegucigalpa), Sept. 26, 1992, p. 1.
2. Cultural geographer Peter H. Herlihy of Southeastern Louisiana University and British development worker Andrew Leake developed the map with twenty-one local "surveyors," who came from twenty-two indigenous population centers in La Mosquitia. MOPAWI and MASTA, a Mosquitia advocacy group, organized the project, which was funded by Cultural

Survival and the Inter-American Foundation. The congress was held in Tegucigalpa's Hotel San Martín, Sept. 22–23, 1992.

3. Schwimmer, "Land use map," p. 4.

4. What will happen to Granja Loma Linda after Elías? How does one turn a good development program into a sustainable one? See Appendix 2 for a discussion of the future of Loma Linda and strategies for building institutions to foster development.

Appendices

Appendix 1

Quotable Quotes by Don José Elías Sanchez

On the Human Condition

Human misery is not lack of money, it's not knowing who you are.

The phrase "poor people" is one of the most offensive expressions. It should be "people who can't see opportunities."

It's impossible to plant ideas of superiority in the brain of someone with an empty stomach.

To make a mistake is not a mistake, but to make a mistake and not admit it—that is a mistake.

Why am I so tough? Because I know Latinos. But I am tough with love.

The body has three parts: The brain is your computer; your hands execute its orders, and your heart empowers your action with love or hate; the rest is just filler.

On Agricultural Development

If the mind of a campesino is a desert, his farm will look like a desert.

"Technology transfer" is an offensive concept; you have to transform people.

The human farm produces, the physical farm reproduces, for any creation exists first in the mind of the architect.

When I started visiting farmers in their fields, many people asked: "Why are you helping us?" I said it was good business for me, because I like to be surrounded by educated people, not ignoramuses!

To see people change is a harvest, to see a man sell his grain and have enough money to buy a cow, that is a harvest, no?

When some Americans come here and see misery, they have a bleeding heart and they give things away. I'm not going to give anything to these guys. Why does a campesino fail? Because he's been handed inputs!

Development is a process of displacement—good ideas displacing bad ones. We don't teach, we share information in two directions.

Spark plugs are the only specialists we need in development.

Agronomists are pests.

On Spiritual Motivation

Christ didn't preach religion, he preached equity, liberty, justice, and love. If you don't love a person, you will never help them.

The world's development has a price—we pay with our souls.

When people begin to develop, only God can escort them.

My philosophy is simple: When I pass away, I'll have enough time to rest.

On the Lessons of Creation

Don't overexpose success. Don't imitate the chicken, imitate the duck. Ducks quietly lay big eggs. Chickens make a lot of noise and lay little ones.

Nature rejoices . . . flowers are the smiles of the soil.

If you want to make progress, then sit down on a hill of fire ants!

On Development Aid

International organizations should act as yeast—they should help national efforts rise.

When I bring bankers to Loma Linda, I show them the importance of sponsoring education: how many ignorant people do you think will open bank accounts?

Granja Loma Linda Methodology of Human Transformation

1. Initiate change in small increments—geographically, technologically, and conceptually.

2. Train by doing: Live with and work alongside farmers.

3. Respect human dignity in action and language.

4. Achieve innovation at minimum cost; use local resources in harmony with nature.

5. Achieve all tasks with excellence—no mediocrity.

6. Share what you have learned; ideas unshared have no value.

7. Create satisfaction, both personal and communal.

8. Innovate based on God's wisdom expressed in the Creation, for the process of lasting change is a spiritual one.

APPENDIX 2

Building Local Institutions
The Case of Granja Loma Linda

WHAT WILL happen to Granja Loma Linda after Elías? How does one turn a good development program into a sustainable one? This case study presents options under discussion for the future of Loma Linda and frameworks to evaluate the institution's sustainability. It was prepared as a basis for group discussion rather than to illustrate either effective or ineffective handling of an administrative situation.

On a sunny Saturday in September 1992, Honduran Don José Elías Sánchez was striding up the terraces of his hillside training farm, Granja Loma Linda , deep in thought. His most recent group of trainees had just begun a long drive back to their homes in Lempira, near the Salvadoran border, and he would have a day and a half to tend to the farm and prepare himself before the next group came. A farm boy who had entered first grade at age fourteen, Don

Prepared by Katie Smith and Finn-Olaf Jones, European Institute of Business Administration (INSEAD), June 1993

Elías found his vocation in helping campesino families improve their farming and their lives. His goal was to enable subsistence farmers, who made up 45 percent of Honduras's population, to provide for themselves on their land, thereby resisting the pull of urban migration and the resulting slum poverty that is now endemic to capital cities of developing countries. His efforts are manifest at the Granja, developed in 1980, which has trained over 30,000 campesinos to date, most of whom stayed on their land, and many of whom launched their own training centers in their respective departments.

For Don Elías, economic development begins with the mind. "Dissatisfaction is the beginning of change," he is fond of saying. And as he walked his farm's terraces that day, bending down from time to time to jerk out a weed or two from between rows of lettuce and onions, he found himself dissatisfied. Now sixty-four and without a relative who could carry on the work of the Granja, Don Elías wondered whether and how he could ensure the training center's survival. He even wondered, with other training farms springing up, whether he *should* ensure its survival. Mulling this over, he returned to his little sleeping hut by the river and placed a favorite country-western cassette in his tape player. With the music of the river and of Charlie Pride, he settled in for a siesta.

Honduran Rural Development

Today, after massive infusions of foreign economic aid and relative peace for more than two decades, Honduras's population of 4.4 million has slipped, by many indicators, to the poorest in Central America. The country's rural sector, the largest in the region (61 percent of its population), counted 77 percent of its members in absolute poverty.[1] In addition, Honduras has the region's highest infant mortality and population growth, lowest life expectancy, and second lowest gross domestic product per capita—just a few dollars more than war-torn Nicaragua. Politicians and bilateral donors acknowledge that the key to improving these statistics lies in improving the resources and resourcefulness of the campesinos.

Bilateral Donors

The largest bilateral donor to Honduras has long been the United States, which since 1985 has provided the country with more than $1 billion of economic aid. The U.S. Agency for International Development (USAID) funds rural development initiatives through the Honduran Ministry of Natural Resources. Historically, USAID has focused its assistance to campesinos on the diffusion of new technologies; however, insiders and observers acknowledge that so far these efforts have failed to produce long-term sustainability.

Nongovernmental Organizations (NGOs)

The first nongovernmental aid agencies in Honduras were largely outgrowths of church ministries to help the poor. By 1972, the Catholic Church in Honduras had launched a village-based Bible study movement—Delegates of the Word—which had a social outreach component that, among other things, advocated that farmers cease burning their fields and actively conserve their topsoil. About the same time, evangelical churches formed a nonprofit called Diaconia, derived from the Greek for "service" and connoting help for physical and spiritual needs. In 1974, Don Elías founded his Association for Coordinating Resources in Development (ACORDE), which was instrumental in matching assistance from international nonprofits to hillside farmers's needs. In 1980, ACORDE brought Oklahoma-based World Neighbors to Honduras; its agricultural training programs were and are considered benchmarks in Central America.

Throughout the 1980s, the NGO movement surged. By 1992, more than 200 NGOs were registered with the Honduran government, many of them focused on food security and using World Neighbors' and Loma Linda techniques, which combine agricultural improvement with environmental conservation. Today, Granja Loma Linda counts among its sponsors for training sessions Catholic Relief Services, World Vision, World Neighbors, the UN Food and Agriculture Organization, the U.S. Peace Corps, and the

Honduran government. They all look to Elías to uncork enthusiasm for change among program beneficiaries. The NGOs then conduct their own follow-up back in the field.

Affiliates of Granja Loma Linda

Granja Loma Linda has spawned other NGOs. Over a dozen former trainees have replicated Elías's concept and built demonstration farms on the Loma Linda model back in their departments. Some of these second-generation farms are supported by international NGOs; others remain purely voluntary efforts in their communities. One of Elías's first protégés, Ismael, now has a thriving demonstration farm in Olancho funded by Davis, California–based Freedom from Hunger Foundation.

By 1992, Elías also had a partnership with the prestigious Pan American Agricultural School, one of Latin America's top agronomy institutes, which had recently received a grant from the Kellogg Foundation to create a rural development institute. The institute sent students to the farm every Friday to learn its techniques, and it hired Elías to facilitate training sessions for beneficiaries of its field projects.

Elías had a special friendship with one graduate of the institute, a young man named Milton Flores. Flores, thirty-five, ran a Honduran-based international NGO: the International Cover Crops Clearinghouse (CIDICCO). Founded in 1988 with grants from the Ford Foundation and the Inter-American Foundation, CIDICCO created an information network on the use of soil-improving cover crops that reached village farmers in more than fifty countries. CIDICCO also facilitated field work and training sessions in Honduras. Flores's work to improve soil picked up where Elías's efforts to conserve it left off. The Flores family often visited Loma Linda to chat and exchange ideas with Elías and Candida. Sometimes Milton and Elías even toyed with the idea of creating a foundation for farmer training that could absorb both Loma Linda and CIDICCO.

Benchmarking the Granja

International Benchmarks. The Granja's methods compare favorably to NGO benchmarks derived from a 1991 symposium on food, security, and the environment held at Arizona State University's School of Agribusiness and Environmental Resources. The symposium collected and debated "lessons learned" on five continents by NGOs working in agricultural development. Entitled "Growing Our Future," the symposium incorporated views from representatives of government agencies, universities, business, conservation groups, and social action nonprofits, and its resolutions were entered in the working papers of the 1992 UN Conference on Environment and Development (the Earth Summit) in Rio de Janeiro. Exhibit 1 lists key factors of success in three NGO projects studied at the conference and compares the Granja's methodology with theirs.

Model for Honduras? According to the 1992 *Honduras environmental agenda:* "There is an urgent need for a rural development model that combines short-term survival measures with long-term ones to restore and conserve the most depleted resources and improve living standards" (p. 68). The agenda specifies Loma Linda as one of the models that could be applied throughout Honduras, given the "required resources for infrastructure development" (p. 69). Soil conservation and small farm development are especially important to stem the current migration of campesinos into Honduras's already swollen urban slums. But Exhibit 2, a matrix for evaluating project sustainability, shows that Granja Loma Linda has some significant vulnerabilities as a model for rural development.

Options for the Future

Elías flipped the Charlie Pride tape and pondered his options for Loma Linda: He might turn over the management to his protégés, although they would have to find help to care for their own training centers. Or he could create a management committee to run the farm and become an umbrella organization for replications. ACORDE, though dormant, is still a legal entity, with a board of directors. He had once considered a long-term partnership with an international NGO, but his dislike of paperwork and love of autonomy made such a partnership difficult.

On the other hand, Elías isn't even certain that his original mission for the farm can endure without the force of his personality and commitment. He recently won a UNESCO medal and became the first non-American to win the World Neighbors Partner Award for his contributions to community development. Perhaps that was a fitting end and he should gradually phase out Loma Linda and allow the techniques he promoted to spread through the large network of campesinos and NGOs now practicing them.

Whatever option Elías chooses, two things are certain: It will be difficult for him to give up the reins of the farm. And it will be even more difficult for someone to follow in his footsteps.

Questions for Discussion

1. What are the keys to the Granja's success, and are they sustainable?
2. Should the government sponsor replication of the Granja as a model for rural development?
3. What should Don Elías do to ensure the future of human farming?

Note

1. Tom Barry and Deborah Preusch, *The Central American Fact Book* (New York: Grove Press, 1986).

Exhibit 1
Lessons Learned in Three International NGO Projects vs.
Granja Loma Linda Methodology

Tuareg Rehabilitation Project, Mali—World Vision	*The Granja*
Enhancement of the resource base should be both a spiritual activity and a physical one for long-term effectiveness.	Loma Linda emphasizes intellectual, physical, and emotional commitment to farming and uses parables to provide motivation for its teachings.
Influence perceptions of appropriate technology first or it won't be used.	The success of the Granja's methods depends on inspiring campesinos to believe in the tools of the human farm—their head, hands, and heart—and on demonstrating organic systems on their farms.
Both community and government participation are essential for effective programming.	Although the Granja gets some government funding, it is essentially a small, independent unit. Government help would be required for replication and geographic distribution of the Granja model.

Sierra de las Minas Biosphere Reserve, Guatemala—World Wildlife Fund	*The Granja*
Start where people are located; focus the project on what they already do.	Follow-up work by Granja staff and NGO sponsors is done at campesinos' farms.
Identify major environmental and economic problems and look for alternative methods, techniques, and organizational structures to address these problems.	Farming techniques are modern yet accessible to all campesinos. Campesinos' social problems (e.g., lack of hygiene, poor self-image) are also addressed to increase campesinos' competence and confidence in resource management.
Induce experimentation on a small scale with these new ideas.	The Granja is a small experimental farm that teaches by example and encourages replication.

(cont.)

EXHIBIT 1 (cont.)

Sierra de las Minas Biosphere Reserve, Guatemala—World Wildlife Fund (cont.)	The Granja
Adapt and perfect the new approaches in local partnerships.	More than a dozen training centers modeled after Loma Linda have been developed in regions of Honduras by former Loma Linda trainees.
Extend practices horizontally by incorporating successful participants as local promoters.	Campesino participants often share their training with their neighbors. However, that is no guarantee that neighbors will imitate them. Some former trainees have met resistance from neighbors to the labor intensity of Loma Linda techniques. Others have been accused of showing off.
Increase the range of activities of the project as the community gains confidence and becomes more innovative.	The Granja has maintained its teaching format but expanded its agricultural techniques and range of trainees. It now trains urban gardeners and agronomists as well as campesinos.

Bolivian Greenhouses Project— Food for the Hungry	The Granja
Promote local ownership of the program.	Elías has resisted the temptation to manage several farms over a wide geographic area, relying instead on replication by locally based teachers.
Demonstration projects should be complemented by extension work.	Granja teaching is complemented by follow-up back at campesinos' farms.
Leave room for creativity in the standard designs.	The Granja has garnered new agricultural techniques from volunteers from around the world as well as from local campesinos.
Poor people value quick results.	Results are immediate in the form of higher yields on campesino homesteads.

Source: Derived from case studies in Katie Smith and Tetsunao Yamamori (eds.), *Growing Our Future: Food Security and the Environment* (West Hartford, Conn.: Kumarian Press, 1992). Used with permission.

Exhibit 2
Capabilities and Vulnerabilities of Granja Loma Linda as a
Model for Honduran Rural Development

Features	Vulnerabilities	Capabilities
Physical/material What productive resources, skills, and hazards exist?	Current farm managers are weak administrators. There would be a problem handling administrative details should the farm and/or farm network expand. The spread of the model would be dependent on receiving resources from a cash-poor local government.	The Granja measures only 14 hectares and has very diverse terrain that is ideal for teaching different cultivation methods. It has no heavy equipment and uses 70% organic materials for cultivation. Meaningful replication of this model would require resources for demonstration farms, scholarships, and specialized studies.
Social/ organizational What are the relations and organization among people?	Elías's charismatic leadership and 37 years of teaching experience are irreplaceable. Long-term success depends on finding and training good, like-minded teachers.	Follow-up visits to campesinos' farms ensure proper implementation of Granja techniques. A network of other NGOs and social workers has sprung up around the Granja in order to replicate the project.
Motivational/ attitudinal How does the community view its ability to change?	Elías's blunt style and domineering presence are sometimes alienating. Not everyone will accept having sense kicked into them. For reasons of pride or defeat, farmers still tend to resist change within their communities.	The farm's success is overwhelming, given that its techniques are replacing destructive, traditional agricultural techniques through a process of education. The holism of this approach has likewise transformed self-destructive social practices.

Source: Based on Mary B. Anderson and Peter Woodrow's model in *Rising from the Ashes: Development Strategies in Times of Disaster* (Boulder: Westview Press, 1989). Used with permission.

Reader's Reference

ACORDE. Association for Coordinating Resources in Development, a not-for-profit Honduran organization founded by Elías Sánchez and Wilmer Dagen in 1974 to match national and international resources for rural development with needs in the country. In 1980, ACORDE invited and arranged for World Neighbors to work in Honduras, and it remained its legal sponsor in the country until 1991.

afforestation. Planting trees to grow forests where historically there has been no forest cover.

agribusiness. Businesses related to the production, transformation, and marketing of agricultural products.

agronomos. Spanish for agronomists.

appropriate technology. Refers to the simplest technology that uses local resources for accomplishing what needs to be done; for example, using animal manure for fertilizer and using oxen or mules for plowing.

ASPTA/Brazil. Portugese acronym for Consultancy and Services for Alternative Agriculture Project, a Brazilian nongovernmental organization.

Bay of Pigs. A bay in west Cuba where anti-Castro Cuban exiles, financed and trained by the U.S. Central Intelligence Agency, landed April 19, 1961, in an attempt to overthrow the Communist regime. Castro subdued the invaders within three days, capturing 1,200. The operation, originally planned by President Eisenhower, was approved by President Kennedy and became a foreign policy debacle.

CADERH. Spanish acronym for the Assessment Center for Developing Human Resources in Honduras, based in Tegucigalpa.

CADESCA. Spanish acronym for the Action Committee to Support Central American Economic and Social Development, based in Panama City, Panama.

campesino. Spanish for peasant farmer.

CARE. Cooperative for Assistance and Relief Everywhere (formerly Cooperative for American Relief Everywhere), based in Atlanta, Georgia. CARE is the world's largest not-for-profit organization in relief and development, with $405 million in annual programming in fifty-three countries.

Catholic Relief Services (CRS). One of the oldest and largest not-for-profit organizations for international relief and development, based in Baltimore, Maryland. CRS was active in efforts to assist Europeans after World War II and today has operations throughout the world.

CATIE. Spanish acronym for the Tropical Agricultural Research and Training Center, based in Turrialba, Costa Rica, with satellite offices throughout Central America.

CEDOH. Spanish acronym for the Honduran Documentation Center.

CIDICCO. Spanish acronym for the International Cover Crops Clearing House, a not-for-profit organization that collects and disseminates information via field demonstrations, newsletters, slide shows, conference presentations, and electronic media on the use of cover crops to improve soils. CIDICCO was founded in 1987 by Milton Flores, with sponsorship from World Neighbors and the Ford Foundation, and is based in Tegucigalpa.

COHDEFOR. Spanish acronym for the Honduran Forest Development Corporation, a parastatal organization set up to manage the protection and use of forest resources in Honduras.

composting. An effective method used by many small farmers and gardeners to produce organic matter necessary to enrich and maintain the "life" of the soil. Compost piles are formed by layering crop residues and organic waste of any kind in a heap and must include green matter such as grass cuttings, dry matter such as leaves, and bacteria-filled matter such as manure to provide the nitrogen, potassium, phosphorous, and bacteria needed to make good, balanced compost.

contour ditch. A narrow ditch dug into a hillside or field on the contour to prevent erosion by catching, channeling, and draining rainwater.

COSECHA. Consulting partnership founded by Roland Bunch and several Central American associates in 1992 to train development organizations worldwide in the methodology of people-centered development. COSECHA, whose acronym is the Spanish word for "harvest," is based in Honduras.

cover crop. Any of a number of plants, such as beans, that creep horizontally and provide leafy ground cover. Cover crops often have the ability to fix nitrogen in the soil, returning the nutrients leeched by heavy rainfall or used by demanding crops such as corn. In addition, cover crops can be turned under the soil after harvest to provide organic material. Cover crops such as the velvet bean, lablab bean, jack bean, and squash are frequently sown between rows of corn.

coyote. Spanish slang for a middleman.

COSUDE. Spanish acronym for Swiss Development Corporation, the Swiss government's official aid organization.

CTN. Spanish acronym for National Technical Cooperation (Cooperación Técnica Nacional), a Honduran human rights organization.

Delegates of the Word. A lay movement of the Catholic Church that promotes the reading of Scripture through weekly meetings in rural homes in regions lacking clergy. In Honduras, communities participating in Delegates of the Word have a strong track record for organizing to promote community development.

development. Both a goal and a process that seeks to achieve the broad objectives of economic equity, social justice, cultural integrity, and ecological sustainability. (From Alan T. White,

Lynne Zeitlin Hale, Yves Renard, and Lafcadio Cortesi, eds., *Collaborative and Community-Based Management of Coral Reefs* [West Hartford, Conn.: Kumarian Press, 1994].)

development program. A program that brings assistance to disadvantaged persons and their environments by helping them to help themselves through training and community organizing. Organizing typically revolves around improving community basics such as primary health care, availability of clean water, agricultural systems, schools, or small-scale credit.

Diaconia. A not-for-profit organization formed in the mid-1960s by Mennonite and Moravian churches and the United Church of Christ—with the help of Agricultural Missions, a dependency of the National Council of Churches—to promote community development. Diaconia, derived from the Greek for "service," connotes help for both physical and spiritual needs.

Don. Honorific title in Spanish to show respect for men of age and stature.

EAP. Spanish acronym for the Pan American Agricultural School.

Earth Summit. Nickname for the June 1992 UN Conference on Environment and Development (UNCED) held in Rio de Janeiro. The conference drew 118 world leaders to negotiate and sign treaties on protecting biodiversity and stemming toxic emissions, as well as to agree on principles for managing forests and preventing desertification. The event drew more than 9,000 journalists, the largest media gathering on record. A parallel "people's summit," called the Global Forum, was held concomitantly in Rio and attracted thousands of representatives from citizen groups.

extensionists. One who takes training out of the classroom to assist people where they live and work.

extension work. Any program that promotes learning out in the field rather than in a formal classroom setting.

FAO. The UN Food and Agriculture Organization, based in Rome. Its goal is to improve the management of the world's agricultural resources and food production, especially in poor countries.

FOPRIDEH. Spanish acronym for the Federation of Private Development Organizations of Honduras, which serves as an umbrella group for NGOs in Honduras.

Granja Loma Linda. Spanish name for Elías Sánchez' farm; in English, Beautiful Little Hillside Farm.

GTZ. German acronym for the German government agency for international development assistance.

Hondureño. Spanish for Honduran.

IICA. Spanish acronym for the Interamerican Institute for Agricultural Cooperation.

INDEVOR. Acronym for the INSEAD International Development Organization, a French not-for-profit agency, which works to help business students apply their skills to international development.

in-row tillage. A method of soil preparation in which only the area to be planted is dug by pick or plowed with oxen or mule; the area in between prepared land is left with grass and weeds. Normally, in Honduran subsistence agriculture, the land is not plowed or dug up; instead, seeds are planted in small holes opened up by a stick. In U.S. agriculture, the whole field is plowed.

INSEAD. French acronym for the European Institute of Business Administration in Fontainebleau, France, Europe's largest master of business administration program.

LUPE. Acronym for Land Use Productivity Enhancement, a $50 million rural development program funded by the USAID and implemented by the Honduran Ministry of Natural Resources.

manzana. A Honduran land measurement equal to 1.7 acres.

MOPAWI. Acronym for Mosquitia Pawisa, which means Development of La Mosquitia in the Miskito language. MOPAWI is a not-for-profit organization serving the development needs of the indigenous peoples of Honduras's La Mosquitia region. It was cofounded by Wilmer Dagen and Osvaldo Munguia.

natural or organic farming. A farming method that uses natural processes and local resources. For example, crop residues, animal manures, and leguminous plants are used to fertilize the soil and build up its content. These measures decrease dependence on commercial products, which can be expensive and toxic (such as pesticides), and result in a healthy, natural crop. Natural farming in its purest form is called "do-nothing" farming. The farmer simply rotates crops so that the residue of one becomes organic material for the next.

nitrogen-fixation. The process by which some plants, with the help of soilborne bacteria, take nitrogen from the air to enhance plant growth and, in the process, leave some in the soil for other plants.

nongovernmental organizations (NGOs) or private voluntary organizations. NGOs cooperate with donor agencies and work on their own to provide project-level services related to the ongoing developmental objectives of donor and recipients.

OXFAM. One of the Western world's oldest nongovernmental organizations, whose roots date back to a movement at Oxford University to prevent world famine.

Pan American Agricultural School. One of the most prestigious and rigorous schools of agricultural studies in Latin America. Founded in 1942 by U.S. banana magnate Sam Zemurray, it offers three-, four-, and five-year postsecondary degrees. It is located in Zamorano, Honduras, and is often referred to as Zamorano.

POSCAE/UNAH. Spanish acronym for the Department of Graduate Studies in Development Economics and Planning at the National Autonomous University of Honduras in Tegucigalpa.

quintals. A measure of weight equal to 100 pounds.

reforestation. The process of planting trees to grow in the place of forests that have been cut down.

relief program. A program that brings assistance to disaster victims and their environment, generally in the form of physical handouts such as food, shelter, clothing, medicine, and tools.

Rodale Farm. A leading international proponent of organic farming and research based in Emmaus, Pennsylvania.

SABER. Acronym for the School of Agribusiness and Environmental Resources at Arizona State University.

SCIDE. Spanish acronym for Service for Inter-American Cooperative Education, the U.S. government agency for which Elías Sánchez worked in Honduras in the late 1950s and 1960s, which carried out educational development programs. SCIDE was set up by the Inter-American Cooperative Administration, a Truman-era precursor to President Kennedy's USAID.

selective harvesting. Technique used in timber operations in which foresters cut only trees selected by age or species within a specified tract of forest.

Sputnik. The name of the Soviet craft that became the first spaceship to orbit the Earth. Chagrined at losing the space race, President Kennedy announced that the United States would be the first to land on the moon and poured financial support into NASA.

Stone Container Corporation. Transnational forest products company based in Chicago, Illinois, that lit a fuse among Honduran environmentalists in 1991–92 when it tried to obtain rights to log millions of acres of pine forest in La Mosquitia and adjoining regions.

sustainable development. Development that meets the needs of the present without compromising the ability of future generations to meet their own needs. (From the 1987 Brundtland Commission.)

terracing. Hillside farming technique by which farmers carve out terraces on the contour, securing or stabilizing the outer edges or sides with grasses or stone walls and walls to prevent soil erosion and to create flat surfaces for cultivation.

UNAH. Spanish acronym for the National Autonomous University of Honduras.

UNESCO. Acronym for the United Nations Educational, Scientific, and Cultural Organization, based in Paris, France.

UNICEF. Acronym for the United Nations Children's Fund. Considered one of the UN's most effective agencies, it is based in New York City and focuses on child health and education.

United Fruit Company (UFC). A U.S. fruit company incorporated in 1899 that pioneered banana plantations in Central America and, in many cases, dictated political fates in the region. UFC president Sam Zemurray founded the Pan American Agricultural School in Honduras. UFC became United

Brands Company in 1970, and in 1990 changed its name to Chiquita Brands International. It is based in Cincinnati, Ohio.

U.S. Agency for International Development (USAID). Created under the Kennedy administration, this agency is responsible for administering U.S. foreign economic and social aid programs.

velvet bean. A leguminous cover crop that has been one of the most successful at improving soil. This bean is used extensively by farmers throughout Honduras as a cover crop grown in combination with corn, based on the experiences of farmers on the north coast of Honduras who developed this practice.

World Neighbors. A relatively small nongovernmental organization (annual budget of less than $3.5 million) based in Oklahoma City, Oklahoma. It operates worldwide and has become a leader in techniques to motivate community participation and to create local impact.

World Vision. One of the world's largest nongovernmental organizations in international development (annual budget over $230 million) and a pioneer in child sponsorship programming.

Zamorano. Term used to refer to a current student or an alumnus of the Pan American Agricultural School.

Bibliography

Publications

Anderson, Mary, and Peter Woodrow. *Rising from the Ashes: Development Strategies in Times of Disaster.* Boulder: Westview Press, 1989.

Barry, Tom, and Deborah Preusch. *The Central American Fact Book.* New York: Grove Press, 1986.

Berry, Wendell. *The Gift of Good Land.* San Francisco: North Point Press, 1981.

Bunch, Roland. *Two Ears of Corn: A People-Based Approach to Agricultural Improvement.* Oklahoma City: World Neighbors, 1984.

EAP. *Escuela Agricola Panamericana 1991–1992 Annual Report.* Tegucigalpa: Lithopress, 1992.

EAP. *Zamorano, 50 Years.* Tegucigalpa: EAP, 1992.

Forss, Kim. *Planning and Evaluation in Aid Organizations.* Stockholm: Economic Research Institute, 1985.

Fukuoka, Masanobu. *The One-Straw Revolution: An Introduction to Natural Farming.* Emmaus, Pa.: Rodale Press, 1978.

Galvez, G., M. Colindres, T. M. Gonzalez, and J. C. Castaldi, "Honduras: Caracterización de los productores de granos básicos." *Temas de seguridad alimentaria no. 7.* Panama City: CADESCA, 1990.

131

Honduras Environmental Agenda 1992. Tegucigalpa: Graficentro Editores, 1992.

Kaimowitz, David, David Erazo, Moisés Mejía, and Aminto Navarro. *Las Organizaciones Privadas de Desarollo y la Tranferencia de Tecnología en el Agro Hondureño.* Tegucigalpa: FOPRIDEH/IICA, March 1992.

Korten, David C. *Getting to the 21st Century: Voluntary Action and the Global Agenda.* West Hartford, Conn.: Kumarian Press, 1990.

Licenciatura en Sociologia, CUEG-UNAH. *Puntos de Vista: Revista de Analisis Politico y Social,* No. 5. Tegucigalpa: CEDOH, July 1992.

Pino, Hugo, and Andrew Thorpe. *Honduras: El Ajuste Estuctural y la Reforma Agraria.* Tegucigalpa: POSCAE-UNAH, 1992.

Posas, Mario. *El movimiento campesino hondureno.* Tegucigalpa: Editorial Guaymuras, 1982.

Schlesinger, Stephen, and Stephen Kinzer. *Bitter Fruit: The Untold Story of the American Coup in Guatemala.* New York: Doubleday, 1982.

Smith, Katie, and Tetsunao Yamamori, eds. *Growing Our Future: Food Security and the Environment.* West Hartford, Conn.: Kumarian Press, 1992.

Thorpe, Andy. "America Central no puede tener democracia con hambre: Las politicas de la reforma agraria en Honduras antes de 1982." *POSCAE-UNAH Documentos de Trabajo,* No. 3. Tegucigalpa: POSCAE-UNAH, May 1991.

UNICEF. *Children in Honduras.* Guatemala City: Dixon Print, 1990.

Interviews in Honduras 1992

NGOs Staff

Roland Bunch, director, COSECHA/Honduras, Valle de Angeles
Miriam Dagen, representative for Mexico, Central America, and the Caribbean, World Neighbors, Tegucigalpa
Wilmer Dagen, cofounder, MOPAWI, Santa Lucia

Rafael Diaz, director, World Neighbors, Tegucigalpa
Milton Flores, director, CIDICCO, Tegucigalpa
Loral Patchen, U.S. Peace Corps volunteer, Intibucá
Daniel Romero, Asociación des Instituciones Evangelicas de
Honduras, San Pedro Sula
Francisco Salinas, agricultural projects director, Catholic Relief
Services, Tegucigalpa
Carlos Zelaya, Food and Agriculture Organization, Tegucigalpa
Raul Zelaya, Patricia Cruz, and Ernesto Palacios, Rural Develop-
ment Institute, Pan American Agricultural School, Zamorano
Roberto Zepeda, extensionist, CIDICCO, El Jute

Government Staff

Luís Alvarez, national technical coordinator, Ministry of Natural
Resources
Alfredo Landaverde, former congressman working on agricultural
modernization law
David Leonard, hillside farming technologies adviser, LUPE
Project

USAID/Honduras Staff

Vincent Cusumano, deputy director of Rural Development
Office
Margaret Harritt, environmental programs officer
Rafael Rosario, director, Natural Resources and Environment
Division

Granja Loma Linda

Marcos Aguilar, trainee, Lempira
Walter Albarenga, extensionist-trainee, Lempira
Jorge Amador, instructor
Juana Cerrato, cook
Juanita Cervantes de Franco, trainee, Lempira

134 / The Human Farm

Armida-Lara Escalante, trainee, Lempira
Antonia Ramos Lainez, trainee, Lempira
Giovanni Lara, extensionist-trainee, Lempira
Camilo Mejía, trainee, Lempira
Mauro Mendizabal, Pan American Agricultural School trainee,
    La Paz, Bolivia
Lucio Menjivar, trainee, Lempira
Evelyn Montega, Pan American Agricultural School trainee,
    Guatemala
Candida Osorio, administrator
Maria-Emilia Rodriguez, trainee, Lempira
Elías Sánchez, founder-director

Farmers (Former Granja Loma Linda Trainees)

Fernando and Vilma Andrade, Linaca
Luís Alonzo Morales, Granja Loma Linda
Jose Benito Ponce and Andrés Ponce, El Jute
Ramón Romero, Tegucigalpa
Florentino, Isai, and Adan Santos, Socorro

Index

Advance Praise for
The Human Farm

The Human Farm is one of the cleverest approaches to people-centered development that I've seen in a long time: practical, human, exciting—a success story amid much pessimism.

—Robin Shell, Vice President
Food for the Hungry International

There is no finer, nor more vital goal than to get the world's inhabitants to live in equilibrium with nature. Katie Smith, in her easy-to-read *The Human Farm*, tells us of the extraordinary contribution to this goal of the Honduran farmer/teacher, Don José Elías Sánchez. In so doing, Smith herself makes an important contribution to "sustainable living."

—Charles de Haes
former Director General
WWF-World Wide Fund for Nature

About the Author

Katie Smith is a management consultant in Toronto. She is a former journalist for the *Wall Street Journal/ Europe* and Montreal *Gazette*, researcher at the Harvard Kennedy School of Government and program coordinator at Food for the Hungry International. Her articles have appeared in the *New York Times*, *TIME* Magazine, the Washington *Post*, and Christianity Today.

Also by Katie Smith

Growing Our Future: Food Security and the Environment, eds. Katie Smith and Tetsunao Yamamori (West Hartford, Conn.: Kumarian Press, 1992)

Future Value: Enterprise and Sustainable Development, eds. Olivier Cadot, Michael Milway, and Katie Smith (Fontainebleau, France: INSEAD Press, 1993)

Kumarian Press
Books for a World
that Works

The Human Farm:
A Tale of Changing Lives
and Changing Lands
Katie Smith

Voices from the Amazon
Binka Le Breton

All Her Paths Are Peace:
Women Pioneers in
Peacemaking
Michael Henderson

Summer in the Balkans:
Laughter and Tears after
Communism
Randall Baker

Kumarian Press
630 Oakwood Avenue,
Suite 119
West Hartford, CT
06110-1529 USA

Inquiries
203-953-0214
Fax 203-953-8579
Orders call toll free
800-289-2664

More Advance Praise for
The Human Farm

The Human Farm brings home some of the key lessons of the 'new breed' of development projects: first and foremost, that successful projects start and end with people.

—Olivier Cadot
Associate Professor
European Institute of
Business Administration

The 800 million people in the Americas face many of the same environmental, agribusiness and human challenges of areas most often quoted by development experts—Africa, Asia and the FSU . . . Our future demands a look at the human aspect of development outlined in this book.

—Eric Thor
Director and Professor
School of Agribusiness and
Environmental Resources
Arizona State University